Land Use and Soil Resources

Ademola K. Braimoh • Paul L.G. Vlek

Editors

Land Use and Soil Resources

 Springer

Ademola K. Braimoh
Global Land Project
Sapporo Nodal Office
Hokkaido University
N9W8 Sapporo
060-0809 Japan

Paul L.G. Vlek
Center for Development Research
University of Bonn
Walter Flex Str. 3
53113 Bonn
Germany

ISBN-978-1-4020-6777-8   e-ISBN-978-1-4020-6778-5

Library of Congress Control Number: 2007941782

Cover Images © 2007 JupiterImages Corporation

Chapter 3 © Royal Swedish Academy of Sciences, Stockholm, Sweden

Printed on acid-free paper.

9 8 7 6 5 4 3 2 1

springer.com

Foreword

Soils are considered as increasingly important in global development issues such as food security, land degradation, and the provision of ecosystem services. *Land Use and Soil Resources* synthesizes scientific knowledge about the impact of different land uses on soils in a manner that resource managers can use it. The book offers contribution to the challenge of food production and soil management as population continues to grow in parts of the world already experiencing food insecurity and shrinking arable land. Improved management on existing arable lands is imperative to guarantee food security for the increasing population.

Food importation is important to augment production in Africa, Asia, and Latin America. Countries in these regions are consequently among the largest net importers of nitrogen, phosphorus, and potassium as well as of virtual water in traded agricultural commodities. Nevertheless, soil-fertility decline and water scarcity persist in many countries in the regions. *Land Use and Soil Resources* offers an explanation on the driving factors of nutrient and water flows across world regions, and the need to factor environmental costs into nutrient and water management.

Irrigation is crucial for crop production in many areas of the world characterized by hydrologic scarcity and variability, but poor irrigation practices often saturate land with salts and render croplands barren in the long run. Salinization and waterlogging constitute a threat to the sustainability of irrigation projects in both developed and developing countries. *Land Use and Soil Resources* combines agronomic and environmental facts in a coherent manner to highlight the conditions for the sustainability of irrigation.

The systemic links between cities and rural areas has always posed a formidable challenge to humankind vis-à-vis producing enough crops to feed the populace, and encroachment of cities on agricultural lands and sensitive ecosystems, amongst other problems. As cities develop as centers of nonagricultural production, they also introduce pollutants into the environment. Soil pollution in most cities is at levels warranting instant and urgent action.

Assessment and management of soil quality for land-use planning is increasingly important due to increasing competition for land among many land uses and the transition from subsistence to market-based farming in many countries. The major challenges include predicting land-use suitability and assessing land-use impacts on soil quality to sustain land productivity. This book presents methods for soil-quality assessment using land-evaluation principles and geospatial information technology.

The rate at which soil organic carbon is lost through cultivation and other disturbances undermines the role of soils in buffering climate change. Besides, soil erosion by water associated with early agriculture is currently the most destructive form of soil degradation with profound effects on rural livelihood and environmental sustainability. *Land Use and Soil Resources* documents the strategies to stem further soil carbon losses. It also highlights the successes and challenges of soil and water conservation measures. Soil management strategies require broader sustainable development policy frameworks for success. In the twenty-first century, soils will become more important as an economic and social resource. Soil is vital for human survival on Earth, but paradoxically our cavalier attitude to this natural resource makes its ecosystems one of the most degraded. The task of disseminating knowledge about soils is extremely urgent. The challenges of soil management vis-à-vis human well-being are presented in a scientifically coherent manner in this book. I count it a great privilege to introduce *Land Use and Soil Resources* at this crucial moment of human history.

Rector, United Nations University, Tokyo   Hans van Ginkel
Under-Secretary-General of the United Nations

Contents

Contributors

Ademola Braimoh is Associate Professor at Hokkaido University and Executive Director of the Global Land Project, Sapporo Nodal Office in Japan. A former research fellow of the United Nations University, Ademola studied Soil Science at the University of Ibadan, Geographical Information Systems and Remote Sensing in the University of Cambridge and holds a Ph.D. in Natural Sciences from the Center for Development Research, University of Bonn. His research interests include land-use–soil quality interaction, and application of geospatial technology for land-change studies and environmental management.

Paul L.G. Vlek is Professor at the University of Bonn and Executive Director of its Center for Development Research. He holds a Ph.D. in Soils and Agronomy. His research interests are sustainable development in the developing world, management of natural resources and ecosystems in a world subjected to global change. Specifically he is concerned with water, soil degradation, and nutrient cycles and means to optimize the allocation of these scarce resources to the benefit of communities and society as a whole.

Henk Breman is a biologist who, after his Ph.D. in Biophysics (Free University, Amsterdam), specialized in environment and rural development, with emphasis on Africa. He was involved in education, research, and development projects, and published papers on fisheries, rangelands, livestock, and crops, in the context of agricultural intensification, desertification control, agroforestry, etc. He lived half of his career in West Africa, the last period as Director of IFDC's Africa Division. He is currently Principal Scientist and Advisor, Agriculture and Environment, CATALIST Project, IFDC–Rwanda

Eric Craswell specialized in soil microbiology and soil fertility at the University of Queensland and was awarded his Ph.D. in 1973. He worked on nitrogen cycling and the efficiency of fertilizers for cereal crops at centers in Queensland, Alabama, and the Philippines. After an extended period in research management in the field of land and water resources, he developed an interest in nutrient and water cycling at the global scale and recently served as Executive Officer of the Global Water System project based at ZEF in Bonn.

Bidjokazo Fofana is Senior Agronomist at the Africa Division of IFDC. He holds a Ph.D. in Weed Science and Agronomy from Justus-Liebig University of Giessen. He coordinates development projects with a focus on natural resource management. His research aims to derive methods for improved land management, to optimize inorganic fertilizer-use efficiency through integrated use with organic matter, and to improve crop yield and animal production.

Jonathan Foley is Professor and the Director of the Center for Sustainability and the Global Environment (SAGE) at the University of Wisconsin, where he is also the Gaylord Nelson Distinguished Professor of Environmental Studies and Atmospheric & Oceanic Sciences. Foley's work focuses on the behavior of complex global environmental systems and their interactions with human societies.

Ulrike Grote studied agricultural economics at the University of Kiel. After receiving her Ph.D. from Kiel in 1994, she worked at the OECD in Paris and the Asian Development Bank in Manila. From 1998 to 2006, Grote worked at the Center for Development Research (ZEF) in Bonn, finishing her habilitation in December 2003. Since October 2006, she is Professor and Director of the Institute for Environmental Economics and World Trade at Leibniz Universität Hannover. Her research focuses on international agricultural trade, environmental and development economics.

Daniel Hillel is Senior Research Scientist of the Goddard Institute for Space Studies at Columbia University and Professor of Plant, Soil, and Environmental Sciences at the University of Massachusetts. He is a world-renowned environmental scientist and hydrologist, and an international authority on sustainable management of land and water resources. Professor Hillel has made an enduring and fundamental contribution to our understanding of soil and water as the keys to global environmental management. His 20-plus books include definitive works on arid-zone ecology, low-volume irrigation, and soil and water physics.

Karl Herweg is Program Coordinator at the Centre for Development and Environment and currently appointed as Adjunct Associate Professor at Mekele University, Ethiopia. He holds a Ph.D. in Geography. Besides his fields of specialization, soil erosion and soil and water conservation, he is concerned with transdisciplinary research and teaching in sustainable land management, impact, and outcome monitoring.

Hans Hurni is Professor at the University of Bern and Director of the Swiss National Centre of Competence in Research (NCCR) North-South, an interinstitutional and transdisciplinary 12-year program involving about 350 scientists worldwide, and including about 100 Ph.D. students from over 40 nations. He holds a Ph.D. focusing on climate change and a Habilitation on soil erosion and conservation. His research interests are sustainable land management and syndrome mitigation, and in particular, soil and water conservation and biodiversity management in highland–lowland contexts, particularly in developing and transition countries.

Hanspeter Liniger is the Coordinator of the global WOCAT (World Overview of Conservation Approaches and Technologies) network for documenting, evaluating, and disseminating sustainable land management practices worldwide. During his 10 years in Kenya he has been focusing on research in improving rainfed agriculture through water conservation as well as in assessing the impact of land use on soil productivity and river flows. He is also involved in teaching and supervising students from Europe as well as developing and transition countries in the field of soil and water conservation.

Abdoulaye Mando is Senior Agronomist at the Africa Division of IFDC. He is Head of the Natural Resource Management Program that implements projects with major objective to develop and disseminate integrated soil fertility management technologies in West Africa. He holds a Ph.D. in Soil Ecology from the University of Wageningen. Before joining IFDC, he was involved in a broad spectrum of agricultural research projects in Burkina Faso, as scientist of national and international institutions.

Peter Marcotullio is currently Visiting Associate Professor of Urban Planning at Columbia University and Adjunct Senior Fellow at the United Nations University Institute of Advanced Studies. From 1999 to 2006 he taught at the University of Tokyo in the Department of Urban Engineering, International Urban and Regional Planning Lab. His research interests span urbanization and development, urban and regional planning, and urban environmental change.

Nick Olejniczak holds master's degrees in Environmental Science, and Life Sciences Communication from the University of Wisconsin–Madison. He is the founder of Third Culture Design, Inc., a Madison-based web development firm, and he frequently lectures on online strategic communication.

Takashi Onishi is a Professor at the Research Center for Advanced Science and Technology, University of Tokyo. He obtained a Ph.D. in Urban Engineering from the same university in 1980. He specializes in city planning and regional planning. His research interests include urban engineering, regional planning, and city planning.

Brigitte Portner is a Research Associate at the Centre for Development and Environment, Institute of Geography, University of Bern. Her M.Sc. in Geography focused on land-use strategies of migrant and nonmigrant households in western Mexico. Her research interests are in rural development, land-use change and social learning processes related to sustainable natural resource management.

Navin Ramankutty is Assistant Professor of Geography at McGill University in Montreal. He studies global land-use and land-cover change, and its environmental consequences, using satellite measurements, ground-based survey data, and computer models of ecosystem processes.

Diego de la Rosa is a Professor of Soil Science at the Spanish Research Council (CSIC), Sevilla, Spain. His research focuses on agroecological approaches for the assessment of soil systems, making application of information technology for developing land-use and management-decision support tools. With special reference to the Mediterranean region, he has conducted numerous studies in the area of agricultural land evaluation and degradation modeling, which have been reported in more than 150 publications. Since 1990, his investigation results are being included into the MicroLEIS system (http://www.microleis.com).

Pete Smith is Professor of Soils and Global Change in the School of Biological Sciences at the University of Aberdeen, Scotland, UK. His research interests include the impacts of global change, for example, land-use and climate change, on soils, and the use of soils, especially in agricultural systems, to mitigate climate change. He is a modeler and investigates global change–soil interactions at scales ranging from sites, through continents, to the globe. He is especially interested in soil carbon and greenhouse gas fluxes from soils and from agriculture.

Ramon Sobral is Soil Survey and Land Evaluation Researcher at Argentina National Institute of Agriculture Research (INTA), Buenos Aires, Argentina. He is devoted to soil classification and land interpretations for rural development and land-use planning. During the 1990s, he designed the Productivity Index applied to the Atlas of Soils of Argentina. He also developed indicators of soil quality and soil health for the Argentine Benchmark Soil Series project.

Lulseged Tamene is Postdoctoral Fellow at the University of Bonn, Center for Development Research. His M.Sc. is on GIS/Remote Sensing Applications for resources management in developing countries and holds a Ph.D. in Ecology and Land Management. His research interests are modeling erosion and sedimentation processes, analyzing land-use/cover changes and their implications on land and water resources, and assessing land management options for sustainable use of water resources. Specifically he is interested in GIS, Cellular Automata, and Multiagent-based models for integrated assessment of water- and soil-degradation processes and the major drivers.

List of Figures

List of Tables

Chapter 1
Impact of Land Use on Soil Resources

Ademola K. Braimoh and Paul L.G. Vlek

The land system is a coupled human–environment system comprising land use, land cover, and terrestrial ecosystems (Global Land Project, 2005). Land use connects humans to the biophysical environment. Conversely, the characteristics and changes in the biophysical environment influence our land-use decision-making. Thus, there is a continuum of states resulting from the interactions between natural (biophysical) and human (social) subsystems of land (Fig. 1.1). Though not always, the dynamics of this continuum generally moves toward increasing human dominance and impact. Thus, mitigating adverse environmental changes requires an improved knowledge of human impact on natural processes of the terrestrial biosphere.

Soil is a basic resource for land use. It is the foundation of all civilizations (Hillel, 1992), serves as a major link between climate and biogeochemical systems (Yaalon, 2000), supports biodiversity, and plays an important role in the ability of ecosystems to provide diverse services necessary for human well-being (Young & Crawford, 2004). Thus, soils must not be neglected in any development endeavor either at local, regional, or global level.

Good soils are not evenly distributed around the world. Depending on parent material, climate, relief, vegetation, and time that determine soil formation, soils have inherent constraints that limit their productivity for various uses. Most soil constraints are not mutually exclusive. For instance, highly acidic soils with aluminum toxicity also have high phosphorus-fixation capacity. The inherent constraints of soils for agricultural production vary widely across regions. For example, erosion hazard, defined as very steep slopes (>30%) or moderately high slope (8–30%) accompanied by a sharp textural contrast within the soil profile, varies from 10% for soils of North Africa and Near East to 20% for soils of Europe. On the other hand, shallowness, the occurrence of rocks close to the soil surface, varies from 11 percent for soils of South and Central America to 23% for soils of North Africa and Near East (FAO, 2000).

Human impact is an additional challenge to the inherent constraints of soils to support life on earth. Soil degradation is largely an anthropogenic process driven by socioeconomic and institutional factors. At the global level, five major human causative factors of soil degradation are overgrazing, deforestation, agricultural mismanagement, fuelwood consumption, and urbanization (UNEP, 2002). Soil degradation often occurs so creepingly to the extent that land managers hardly

A.K. Braimoh and P.L.G. Vlek (eds.), *Land Use and Soil Resources.*
© Springer Science+Business Media B.V. 2008

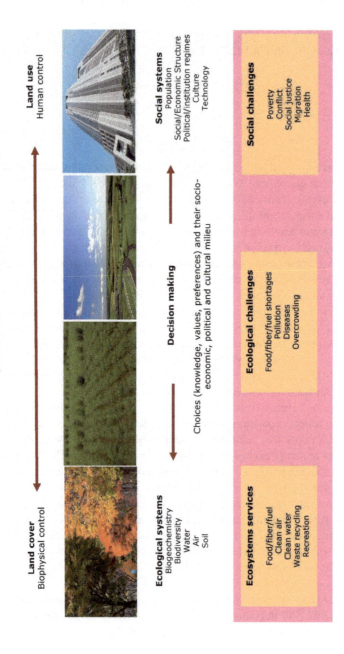

Fig. 1.1 The continuum of states resulting from the interactions between natural and social dynamics. (Adapted from Global Land Project, 2005). Human decisions lead to different states along the continuum from wilderness to mega-cities. Different system characteristics are observed depending on whether biophysical or human control is more dominant. Different ecological and social challenges also result from human interference

contemplate initiating timely ameliorative or counterbalance measures (Vlek, 2005). It is often associated with food insecurity, decline in living standard, social upheavals, and the collapse of civilizations (Weiss et al., 1993; Wood, Sebastian, & Scherr, 2000).

The past few years have witnessed considerable interest in land-use research as a result of the realization of the influence of land-use and land-cover changes on Earth System functioning. Land use still faces large-scale changes and modifications in the near future due to population growth, political and socioeconomic changes related to globalization, and changes in land-related policies. Like the inherent biophysical constraints, human-induced changes on soil ecosystems are highly complex, vary across the world, and are profoundly impacting ecosystem services and human well-being. Though not all human impacts are negative, the environmental crises associated with adverse land-use changes have created a compelling need for a unifying volume that addresses the multifaceted impacts of land use on soils.

This book synthesizes information on the impact of various uses on soils. It is written with scientific clarity to inform policies for sustainable soil management. In Chapter 2, Smith writes on soil organic carbon (SOC) dynamics and management. SOC dynamics, the link between climate and biogeochemical systems, is a major pathway to understanding and predicting human impacts on the Earth (Yaalon, 2000). SOC losses arise from converting grasslands, forests, or other native ecosystems to croplands, or by draining or cultivating organic soils. On the other hand, positive impacts on SOC arise from restoring grasslands, forests, or native vegetation on former croplands, or by restoring organic soils to their native condition. With growing demand for food, more land is required to produce crops, implying greater potentials for SOC losses. Globally, the carbon sink capacity of agricultural and degraded soils is about 50–60% of historical carbon loss (Lal, 2004). While the rate of soil carbon sequestration depends on soil properties, climate, and farming practices, Smith argues for a broader sustainable management framework for the adoption of successful soil carbon management in developing countries. This includes policies to encourage fair trade, reduced subsidies for agriculture in developed countries, and less onerous interest on loans and foreign debts.

Recent estimates indicate that 854 million people are undernourished worldwide, with the highest proportion (about 61%) residing in Asia and the Pacific (FAO, 2003). Sub-Saharan Africa where 10% of the world's human population resides has the highest prevalence of undernourished people of 33%. Food production in many developing countries is hampered by decreasing per capita cropland, soil nutrient depletion, lack of access to intensification inputs, and lack of enabling policy environments that favor smallholders. In Chapter 3, Ramankutty, Foley, and Olejniczak review the major changes in global distribution of croplands during the twentieth century. Between 1900 and 1990, per capita cropland area decreased from 0.75 ha to 0.35 ha—less than the minimum 0.5 ha required to provide an adequate diet (Lal, 1989). Population growth was not met by a corresponding increase in cropland expansion first because increases in food production were achieved by agricultural intensification, and also because cropland expansion did not always occur in the regions with the highest population growth. Thus, food security by nations with insufficient agricultural production was achieved primarily through food aid and trade. As cropland and settlement expansion has claimed prime agricultural land, there is increasing reliance on technology for improved agricultural productivity. Even though food production is generally adequate at the global level, the global food production system is becoming increasingly vulnerable owing to the dependence on technology to increase productivity.

Since the transition from hunting and foraging to agriculture about 10,000 years ago, human dependence and impact on soils have become more apparent, with soil erosion standing out as the most destructive form of soil degradation. Three epochs of soil erosion can be identified since cultivation began in the Fertile Crescent in the Middle East (McNeill & Winiwarter, 2004). The first epoch occurred in the second millennium BC during the expansion of early river basin civilizations, when farmers cleared forested slopes for agriculture. In the next 3,000 years, conversion of forests to farmlands also occurred in Africa, Eurasia and the Americas. The second epoch occurred during the sixteenth to nineteenth century when the introduction of agricultural machinery (plowshares) in Eurasia, North America, and South America accelerated soil erosion on farmers' fields. The third epoch occurred after 1945 when rapid population growth amongst other factors encouraged migration into tropical rainforests. On the average, the soils of the world have lost 25.3 million tons of humus per year since 10,000 years ago. Over the last 300 years the average loss was 300 million tons per year. The last 50 years in particular have brought human-induced soil resources degradation to exceptionally high levels. On the average, 760 million tons of humus has been lost per year in the last half-century (Rozanov et al., 1990). Ethiopia in East Africa is among the countries with the highest soil erosion rates. Its highlands lose over 1.5 billion tons of topsoil per year, leading to a reduction of about 1.5 million tons of grain in the country's annual harvest (Taddese, 2001). While there may be some uncertainties in these statistics, the magnitude of soil erosion problem is largely indubitable.

It is noteworthy that from the 1930s, modern soil conservation endeavors (mostly sponsored by governments) has broadened significantly (McNeill & Winiwarter, 2004), utilizing several techniques, including contour plowing, use of cover crops, and conservation tillage. In Chapter 4, Hurni et al. highlight the development of agriculture since 1950, and elaborate on progress in soil and water conservation techniques. They reveal that social factors (land tenure security, market access, and increased level of participation in decision-making) are necessarily involved in soil water conservation. While measures developed for modern agricultural systems have begun to show positive impacts, external support in the form of investment in sustainable land-management technologies is still required for small-scale farming. In an empirical study reported in Chapter 5, Tamene and Vlek applied soil erosion models to identify high-risk areas to target management intervention in Ethiopia. The model generally predicted higher erosion than deposition, implying that soil loss is higher than the amount that can be redistributed within the catchments, thereby increasing the potential for sediment export. The study further indicates that a land-use planning approach could help reduce erosion problems in Ethiopia, and other parts of the world with similar environmental conditions.

Degradation of soils and the depletion of water resources that caused the collapse of irrigation-based societies about 6,000 years ago are threatening the viability of irrigation at present. The fact that irrigation is vital for increased productivity is well appreciated by farmers, governments, and international donors. However, the expansion of irrigation, which had been a principal focus of agricultural development for the past few years, has lately been offset by the abandonment of older irrigated areas due to depletion of groundwater reserves, waterlogging and salination, or diversion of water supplies to alternative uses. Chapter 6 by Vlek, Hillel, and Braimoh focuses on the prerequisites of sustainable irrigation. The authors explain the process of waterlogging and salt buildup and factors that accentuate the problems. Case studies on how irrigation problems manifest in different parts of the world were also reviewed. The case studies generally indicate the importance of early warning systems to detect the onset of problems in irrigated agriculture. The prospects of climate change further calls for adroit management of irrigation as well as proactive environmental policies. The continuous diminution of good-quality water for irrigation calls for stepping up research to produce crop varieties that require less water and nutrients and have increased salt tolerance. There is also the need to develop economic incentives that encourage water conservation.

The impact of globalization on nutrient and water flows is the focus of Chapter 7 written by Grote, Craswell, and Vlek. The differences between the nutrient and water balance in nutrient- and water-deficit and surplus countries largely reflect the large disparities in resources and agricultural policies between less developed and industrialized countries, respectively. The international net flows of nitrogen, phosphorus, potassium (NPK) nutrients amounts to about $5\,Tg$ ($1\,Tg = 10^{12}\,g$) in 1997 and are projected to increase to about $9\,Tg$ in 2020. This represents a major human-induced perturbation of global nutrient cycles. The major net importers of NPK and virtual water in traded agricultural commodities are West Asia/North Africa and sub-Saharan Africa. Countries with a net loss of NPK and virtual water in agricultural commodities are the major food exporting countries—the USA, Australia, Canada, and Latin America. Agricultural trade liberalization and the reduction of subsidies could reduce excessive nutrient and water use in nutrient- and water-surplus countries and possibly make inputs more affordable to farmers in nutrient- and water-deficient countries. Institutional strengthening and infrastructure development are also required in nutrient- and water-deficient countries. Grote, Craswell, and Vlek also advocate factoring environmental costs into nutrient and water management.

Despite the fact that sub-Saharan Africa is a major net importer of nutrients, the problem of soil-nutrient depletion still persists in the region. Soil-nutrient depletion is one of the major causes of food crises in sub-Saharan Africa. Opinions are however diverse as to the causes of the depletion in the world's most ancient landmass. In Chapter 8, Breman, Fofana, and Mando write on strategies for ameliorating nutrient deficiencies in soils of sub-Saharan Africa. They state that farmers in sub-Saharan Africa deplete soils primarily because the soils are poor by nature. The extremely poor resource base, unfavorable value–cost ratio, and inadequate socioeconomic and policy environments caused the green revolution to bypass Africa. Breman, Fofana, and Mando further explain how the redistribution of organic matter and the nutrients it contains can help in transitioning agriculture in sub-Saharan Africa from extensive to intensive phase.

The concept of soil quality experienced the most rapid adoption in the 1990s as a consequence of the effects of land use on the dynamic aspects of soil quality indicators (Karlen, 2004). Soil quality is a notion that is much broader, but includes the capacity of the soil to supply nutrients, maintain suitable biotic habitat, and resist degradation. Soil quality is the key to agricultural productivity; especially in low-input production systems where productivity-enhancing technologies are largely out of reach of the farmers. Soil quality is not often considered a policy objective by policymakers unless soil degradation threatens other development objectives (Scherr, 1999). The decline in long-term productivity currently constitutes a threat to livelihood in many developing countries, necessitating the development of indicators for soil quality management. Methods to assess soil quality are discussed by de la Rosa and Sobral in Chapter 9. Acknowledging that the task of assessing soil quality is complicated, the authors nonetheless argue for an approach based on knowledge derived from agroecological land evaluation. They also make a case for the incorporation of spatial information technology in soil quality prediction. General trends in soil quality management strategies that can be adapted to different farming situations are discussed.

Urban sprawl is a ubiquitous phenomenon in developed and developing countries. Globally, urban land-use activities potentially remove about 7% (2.4 million km^2) of all cultivated systems from agricultural production, of which a proportion is high-quality farmland (McGranahan et al., 2005). As the world continues to urbanize, we have lost contact with soils and the services they provide to sustain life. In Chapter 10, Marcotullio, Braimoh, and Onishi provide a review of the multiscale impact of cities and urban processes on soils. Though urbanization is not accompanied by economic growth in developing countries, soil pollutant contamination in their cities continues to increase to levels warranting immediate action. The authors argue for a global assessment of urban soils to identify the patterns and processes of anthropogenic impacts. This should help in developing appropriate intervention measures.

Acknowledgments Most of the chapters in this book were presented at the LUCC's[1] Technical Session "Impact of Land-use Change on Soil Resources" at the 6th International Human Dimensions Program on Global Environmental Change (IHDP) Open Meeting in Bonn from 6 to 9 October 2005. We are grateful for the financial assistance provided by the IHDP, the United Nations University Institute of Advanced Studies, and the Japan Ministry of Science and Technology (MEXT) through the Special Funds for Promoting Science and Technology. We thank all authors for their thoughtful contributions, despite their busy schedules. We also thank Chris Barrow, Donald Davidson, Xixi Lu, Anita Veihe, Willy Verheye and Barbara Wick for their constructive reviews.

---

[1] Land Use and Land Cover Change (LUCC) was a joint project of the IHDP and IGBP. LUCC and Global Change and Terrestrial Ecosystems (GCTE) have been succeeded by the Global Land Project (www.globallandproject.org).

References

FAO (Food and Agriculture Organization of the United Nations) (2000). *Land resource potential and constraints at regional and country levels*. Rome: FAO, 122 pp.

FAO (Food and Agriculture Organization of the United Nations) (2003). *The state of food insecurity in the world 2003*. Rome: FAO.

Global Land Project (2005). *Science plan and implementation strategy*. IGBP Report No. 53/ IHDP Report No. 19, IGBP Secretariat, Stockholm, Sweden.

Hillel, D. (1992). *Out of the Earth: Civilization and the life of the soil*. Berkeley, CA: University of California Press.

Karlen, D. (2004). Soil quality as an indicator of sustainable tillage practices. *Soil Tillage Research, 78*(2), 129–130.

Lal, R. (1989). Land degradation and its impact on food and other resources, In D. Pimentel & C. W. Hall (Eds.), *Food and natural resources* (pp. 85–140). San Diego, CA: Academic Press.

Lal, R. (2004). Soil carbon sequestration impacts on global climate change and food security. *Science, 304*, 1623–1627.

McGranahan G., Marcotullio P. J., et al. (2005). Urban systems. In Millennium Ecosystem Assessment (Ed.), *Current state and trends: Findings of the condition and trends working group. Ecosystems and human well-being, Vol. 1* (pp. 795–825). Washington DC: Island Press.

McNeill, J. R., & Winiwarter, V. (2004). Breaking the sod: Humankind, history, and soil. *Science, 304*(5677), 1627–1629.

Rozanov B. G., Targulian V., & Orlov D. S. (1990). Soils. In B. L. T. II., W. C Clark, R. W Kates, J. F Richards, J. T Mathews, & W. B Meyer (Eds.), *The Earth as transformed by human action, global and regional changes in the biosphere over the past 300 years* (pp. 203–214). Cambridge: Cambridge University Press.

Scherr, S. J. (1999). *Soil degradation a threat to developing country food security by 2020*? Food, Agriculture, and the Environment Discussion Paper 27. Washington DC: International Food Policy Research Institute.

Taddese, G. (2001). Land degradation: A challenge to Ethiopia. *Environmental Management, 27*, 815–824.

UNEP, United Nations Environment Programme. (2002). *Global environment outlook 3. Past, present and future perspectives*. London: Earthscan.

Vlek, P. L. G. (2005). *Nothing begets nothing: The creeping disaster of land degradation*. Bonn, Germany: Interdisciplinary Security Connections Publications Series of United Nations University Institute of Environment and Security (28 pp.).

Weiss, H., Courty, M. A., Wetterstrom, W., Guichard, F., Senior, L., Meadow, R., & Curnow, A. (1993). The genesis and collapse of third millennium north Mesopotamian civilization. *Science, 261*(5124), 995–1004.

Wood, S., Sebastian, K., & Scherr, S. J. (2000). *Pilot analysis of global ecosystems: Agroecosystems*. Washington DC: International Food Policy Research Institute.

Yaalon, D. H. (2000). Down to earth: Why soil and soil science matters. *Nature, 407*, 301.

Young I. M., & Crawford, J. W. (2004). Interactions and self-organization in the soil-microbe complex. *Science, 304*(5677), 1634–1637.

Chapter 2
Soil Organic Carbon Dynamics and Land-Use Change

Pete Smith

Abstract Soils contain more than twice the amount of carbon found in the atmosphere. Historically, soils have lost 40–90 Pg C globally through cultivation and disturbance. Current rates of carbon loss due to land-use change are about 1.6 ± 0.8 Pg C y^{-1}, mainly in the tropics. The most effective mechanism for soil carbon management would be to halt land-use conversion, but with a growing population in the developing world, and changing diets, more land is likely to be required for agriculture.

Maximizing the productivity of existing agricultural land and applying best management practices to that land would slow the loss of, or is some cases restore, soil carbon. However, there are many barriers to implementing best management practices, the most significant of which in developing countries are driven by poverty and in some areas exacerbated by a growing population. Management practices that also improve food security and profitability are most likely to be adopted. Soil carbon management needs to be considered within a broader framework of sustainable development. Policies to encourage fair trade, reduced subsidies for agriculture in developed countries, and less onerous interest on loans and foreign debt would encourage sustainable development, which in turn would encourage the adoption of successful soil carbon management in developing countries.

Keywords Soil organic carbon (SOC), land-use change, sequestration, barriers, sustainable development, climate mitigation

2.1 Introduction

2.1.1 Soils and the Global Carbon Cycle

Globally, soils contain about 1,500 Pg (1 Pg = 1 Gt = 10^{15} g) of organic carbon (Batjes, 1996), about three times the amount of carbon in vegetation and twice the amount in the atmosphere (IPCC, 2000a).

A.K. Braimoh and P.L.G. Vlek (eds.), *Land Use and Soil Resources.*
© Springer Science + Business Media B.V. 2008

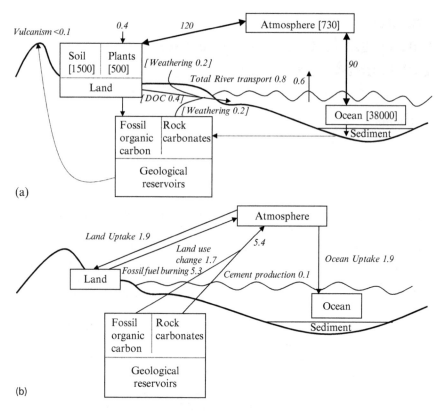

Fig. 2.1 The global carbon cycle for the 1990s (Pg C). **a** The natural carbon cycle (DOC = dissolved organic carbon). **b** The human perturbation (redrawn from IPCC, 2001; Smith, 2004)

The annual fluxes of CO_2 from atmosphere to land (global Net Primary Productivity [NPP]) and land to atmosphere (respiration and fire) are each of the order of 60 Pg C y^{-1} (IPCC, 2000a). During the 1990s, fossil fuel combustion and cement production emitted 6.3 ± 1.3 Pg C y^{-1} to the atmosphere, whilst land-use change emitted 1.6 ± 0.8 Pg C y^{-1} (Schimel et al., 2001; IPCC, 2001). Atmospheric carbon increased at a rate of 3.2 ± 0.1 Pg C y^{-1}, the oceans absorbed 2.3 ± 0.8 Pg C y^{-1} with an estimated terrestrial sink of 2.3 ± 1.3 Pg C y^{-1} (Schimel et al., 2001; IPCC, 2001).

The size of the pool of soil organic carbon (SOC) is therefore large compared to gross and net annual fluxes of carbon to and from the terrestrial biosphere. Figure 2.1 (IPCC, 2001) shows a schematic diagram of the carbon cycle.

2.1.2 Factors Controlling SOC Levels

The level of SOC in a particular soil is determined by many factors including climatic factors (e.g., temperature and moisture regime) and edaphic factors (e.g., soil parent material, clay content, cation-exchange capacity). For a given soil type, however, SOC stock can also vary, the stock being determined by the balance of net carbon inputs to the soil (as organic matter) and net losses of carbon from the soil (as carbon dioxide, dissolved organic carbon, and loss through erosion). Carbon inputs to the soil are largely determined by the land use, with forest systems tending to have the largest input of carbon to the soil (inputs all year round) and often this material is also the most recalcitrant. Grasslands also tend to have large inputs, though the material is often less recalcitrant than forest litter and the smallest input of carbon is often found in croplands which have inputs only when there is a crop growing and where the carbon inputs are among the most labile. The smaller input of carbon to the soil in croplands also results from removal of biomass in the harvested products, and can be further exacerbated by crop residue removal, by tillage which increases SOC loss by breaking open aggregates to expose protected organic carbon to weathering and microbial breakdown, and also by changing the temperature regime of the soil. Impacts of land-use change on SOC are discussed further in Section 2.2.

2.1.3   Historical and Current Losses of SOC due to Land-Use Change

Soil carbon pools are smaller now than they were before human intervention. Historically, soils have lost between 40 and 90 Pg C globally through cultivation and disturbance (Houghton, 1999; Houghton et al., 1999; Schimel, 1995; Lal, 1999). It is estimated that land-use change emitted 1.6 ± 0.8 Pg C y^{-1} to the atmosphere during the 1990s (IPCC, 2001; Schimel et al. 2001).

2.2   Land-Use Change and SOC Loss

In a recent modeling study examining the potential impacts of climate and land-use change on SOC stocks in Europe, land-use change was found to have a larger net effect on SOC storage than projected climate change (Smith et al., 2005a). Indeed, most long-term experiments on land-use change show significant changes in SOC (e.g., Smith et al., 1997, 2000, 2001a, 2002).

Guo and Gifford (2002) conducted a meta-analysis of land-use change experiments and showed that converting forestland or grassland to croplands caused significant loss of SOC, whereas conversion of forestry to grassland did not result is SOC loss in all cases, though total ecosystem carbon presumably decreased due to loss of the tree biomass carbon. Similar results have been found in Brazil, where total ecosystem losses are large but the soil carbon does not decrease (Veldkamp, 1994; Moraes et al., 1995; Neill et al., 1997; Smith et al., 1999), though other studies have shown a loss of SOC upon conversion of forest to grassland (e.g., Allen, 1985;

Mann, 1986; Detwiller & Hall, 1988). In the most favorable case, only about 10% of the total ecosystem carbon lost after deforestation (due to tree removal, burning, etc.) can be recovered (Fearnside, 1997; Neill et al., 1997; Smith et al., 1999).

The largest per-area losses of SOC occur when organic soils (e.g., peatlands) are drained, cultivated, or limed. Organic soils hold enormous quantities of SOC, accounting for 329–525 Pg C, or 15–35% of the total terrestrial carbon (Maltby & Immirizi, 1993), with about one-fifth (70 Pg) located in the tropics. Studies of cultivated peats in Europe show that cultivated organic soils can lose significant amounts of SOC through oxidation and subsidence; between 0.8 and 8.3 t C ha^{-1} y^{-1} (Nykänen et al., 1995; Maljanen et al., 2001, 2004; Lohila et al., 2004;). The potential for SOC loss from land-use change on highly organic soils is therefore very large.

In short, negative impacts on SOC arise from converting grasslands, forests, or other native ecosystems to croplands, or by draining, cultivating, or liming highly organic soils. Positive impacts on SOC arise from restoring grasslands, forests, or native vegetation on former croplands, or by restoring organic soils to their native condition. On managed land, best management practices that increase carbon inputs to the soil (e.g., improved residue and manure management) or reduce losses (e.g., reduced impact tillage, reduced residue removal) help to maintain or increase SOC levels. Management practices to increase SOC storage are discussed in Section 2.3.

The most effective mechanism to reduce SOC losses would be to halt land conversion to agriculture, but with population growing and diets changing in developing countries (Smith et al., 2006b), more land is likely to be required for agriculture. To meet growing and changing food demands without encouraging land conversion to agriculture will require productivity on current agricultural land to be increased. In addition to increasing agricultural productivity, there are a number of other management practices that can be used to prevent SOC loss. These are described in more detail in Section 2.3.

2.3 Land-Use Change and Land Management to Restore/Sequester SOC

2.3.1 Global Potential for Soil Carbon Sequestration

Soil carbon sequestration can be achieved by increasing the net flux of carbon from the atmosphere to the terrestrial biosphere by increasing global NPP (thus increasing carbon inputs to the soil), by storing a larger proportion of the carbon from NPP in the longer-term carbon pools in the soil, or by slowing decomposition. For soil carbon sinks, the best options are to increase carbon stocks in soils that have been depleted in carbon, that is, agricultural soils and degraded soils (see Section 2.2).

Estimates of the potential for additional soil carbon sequestration vary widely. Based on studies in European cropland (Smith et al., 2000), US cropland (Lal et al.,

1998), global degraded lands (Lal, 2001), and global estimates (Cole et al., 1996; IPCC, 2000a), an estimate of global soil carbon sequestration potential is 0.9 ± 0.3 Pg C y^{-1} was made by Lal (2004a, 2004b), between a one-third and one-fourth of the annual increase in atmospheric carbon levels. Over 50 years, the level of carbon sequestration suggested by Lal (2004a) would restore a large part of the carbon lost from soils historically.

The most recent estimate (Smith et al., 2007a) is that the technical potential for SOC sequestration globally is around 1.3 Pg C y^{-1} but this is very unlikely to be realized. Economic potentials for SOC sequestration estimated by Smith et al. (2007a) were 0.4, 0.6, and 0.7 Pg C y^{-1} at carbon prices of US$0–20, US$0–50, and US$0–100 tons CO_2-equavalent^{-1}, respectively. At reasonable carbon prices, then, global soil carbon sequestration seems to be limited to around 0.4–0.7 Pg C y^{-1}. The estimates for carbon sequestration potential in soils are of the same order as for forest trees, which could sequester between about 1 and 2 Pg C y^{-1} (Trexler, 1988 [cited in Metting, Smith, & Amthor 1999]; IPCC, 1996).

Many reviews have been published recently discussing options available for soil carbon sequestration and mitigation potentials (e.g., Lal et al., 1998; Metting et al., 1999; Nabuurs et al., 1999; Follett et al., 2000; Smith et al., 2000; IPCC, 2000a; Cannell, 2003; Freibauer et al., 2004; Lal, 2004a; Smith et al., 2007a). Table 2.1 summarizes the main soil carbon sequestration options available.

Most of the estimates for the sequestration potential of activities listed in Table 2.1 range from about 0.3 to 0.8 t C ha^{-1} y^{-1}, but some estimates are outside this range (Nabuurs et al., 1999; Follett et al., 2000; IPCC, 2000a; Smith et al., 2000, 2007a; Lal, 2004a;). When considering soil carbon sequestration options, it is important also to consider other side effects, including the emission of other greenhouse gases. Smith et al. (2001b, 2007a) showed that as much as one-half of the climate mitigation potential of some carbon sequestration options could be lost when increased emissions of other greenhouse gases (nitrous oxide [N_2O] and methane [CH_4]) were included, and Robertson et al. (2000) have shown that some practices that are beneficial for SOC sequestration, may not be beneficial when all greenhouse gases are considered.

One also needs to consider the trade-off between different sources of carbon dioxide. For example, nitrogen fertilizer production has an associated carbon cost, and some authors have argued that the additional carbon sequestration for increased production is outweighed by the carbon cost in producing the fertilizer (Schlesinger, 1999). However, other studies in developing countries suggest that when accounting for increased production per unit of land allowed by increased fertilizer use, and the consequent avoided use of new land for agriculture, that there is a significant carbon benefit associated with increased fertilizer use in these countries (Vlek et al., 2004).

Soil carbon sinks resulting from sequestration activities are not permanent and will continue only for as long as appropriate management practices are maintained. If a land-management or land-use change is reversed, the carbon accumulated will be lost, usually more rapidly than it was accumulated (Smith et al., 1996). For the greatest potential of soil carbon sequestration to be realized, new carbon sinks, once

Table 2.1 Soil carbon sequestration practices and the mechanisms by which they increase SOC levels

Activity	Practice	Specific management change	Increase C inputs	Decrease C losses	Reduce disturbance
Cropland management	Agronomy	Increased productivity	X		
		Rotations	X		
		Catch crops	X		
		Less fallow	X		
		More legumes	X		
		Deintensification			X
		Improved cultivars	X		
	Nutrient management	Fertilizer placement	X		
		Fertilizer timing	X		
	Tillage/residue management	Reduced tillage			X
		Zero tillage			X
		Reduced residue removal	X		X
		Reduced residue burning	X		X
	Upland water management	Irrigation	X		
		Drainage	X		
	Set-aside and land-use change	Set aside	X		X
		Wetlands	X	X	
	Agroforestry	Tree crops inc. Shelterbelts etc.	X		X
Grazing land management	Livestock grazing intensity	Livestock grazing intensity		X	
	Fertilization	Fertilization	X		
	Fire management	Fire management		X	
	Species introduction	Species introduction	X		
	More legumes	More legumes	X		
	Increased productivity	Increased productivity	X		
Organic soils	Restoration	Rewetting/ abandonment		X	X
Degraded lands	Restoration	Restoration	X	X	X

established, need to be preserved in perpetuity. Within the Kyoto Protocol, mechanisms have been suggested to provide disincentives for sink reversal that is, when land is entered into the Kyoto process it has to continue to be accounted for and any sink reversal will result in a loss of carbon credits.

Soil carbon sinks increase most rapidly soon after a carbon enhancing land-management change has been implemented, but soil carbon levels may decrease initially if there is significant disturbance (e.g., when land is afforested). Sink strength, that is, the rate at which carbon is removed from the atmosphere, in soil becomes smaller with time, as the soil carbon stock approaches a new equilibrium. At equilibrium, the sink has saturated—the carbon stock may have increased, but the sink strength has decreased to zero (Smith, 2004).

The time taken for sink saturation (i.e., new equilibrium) to occur is highly variable. The period for soils in a temperate location to reach a new equilibrium after a land-use change is around 100 years (Jenkinson, 1988; Smith et al., 1996) but tropical soils may reach equilibrium more quickly. Soils in boreal regions may take centuries to approach a new equilibrium. As a compromise, current Intergovernmental Panel on Climate Change (IPCC) good practice guidelines for greenhouse gas inventories use a figure of 20 years for soil carbon to approach a new equilibrium (IPCC, 1997; Paustian et al., 1997).

2.3.2 Soil Carbon Sequestration to Help Meet Atmospheric CO_2 Concentration Stabilization Targets

The current annual emission of CO_2-carbon to the atmosphere is $6.3 \pm 1.3\,\mathrm{Pg}$ C y^{-1}. Carbon emission gaps by 2100 could be as high as $25\,\mathrm{Pg}$ C y^{-1} meaning that the carbon emission problem could be up to four times greater than at present. The maximum annual global carbon sequestration potential $= 0.9 \pm 0.3\,\mathrm{Pg}$ C y^{-1} meaning that even if these rates could be maintained until 2100, soil carbon sequestration would contribute a maximum of 2–5% toward reducing the carbon emission gap under the highest emission scenarios. When we also consider the limited duration of carbon sequestration options in removing carbon from the atmosphere, we see that carbon sequestration could play only a minor role in closing the emission gap by 2100. It is clear from these figures that if we wish to stabilize atmospheric CO_2 concentrations by 2100, the increased global population and its increased energy demand can only be supported if there is a large-scale switch to non-carbon-emitting technologies for producing energy.

Given that soil carbon sequestration can play only a minor role in closing the carbon emission gap by 2100, is there any role for carbon sequestration in climate mitigation in the future? The answer is yes. If atmospheric CO_2 levels are to be stabilized at reasonable concentrations by 2100 (e.g., 450–650 ppm), drastic reductions in emissions are required over the next 20–30 years (IPCC, 2000b). During this critical period, all measures to reduce net carbon emissions to the atmosphere would play an important role—there will be no single solution (IPCC, 2000b). Given that carbon sequestration is likely to be most effective in its first 20 years of implementation, it should form a central role in any portfolio of measures to reduce atmospheric CO_2 concentrations over the next 20–30 years whilst new energy technologies are developed and implemented (Smith, 2004).

2.4   Overcoming Barriers to Implementing Best Management Practices for Enhancing SOC Stocks

There are a number of barriers that may prevent best management practices for soil carbon being implemented. These fall into five categories—economic, risk-related, political/bureaucratic, logistical, and educational/societal barriers. Some of the most important barriers to implementation are given in Table 2.2. Trines et al. (2006) considered barriers preventing a range of agricultural and forestry green-house gas mitigation measures (including soil carbon sequestration) in developed countries, developing countries, and countries with economies in transition.

Transaction and monitoring costs can be barriers in all regions of the world, but other economic barriers, such as cost of land, are important barriers mostly in developing countries and countries with economies in transition, even though some landowners in industrialized parts of the world with high population densities may argue that the establishment of shopping malls and condominium developments is economically more appealing.

In developing countries, continued poverty, lack of existing capacity, and population growth continue to prevent application of management practices that optimize yields and profits, before even considering mitigation. Because of continued population growth in developing countries, competition from other land use is a barrier to implementation. In developed countries, competition from other land uses is a serious barrier, in particular for dedicated bioenergy crop production and practices that require agriculture to be abandoned in particular areas, such as

Table 2.2   Main barriers to implanting successful soil carbon management strategies

Broad category of barrier	Barrier
Economic	Cost of land
	Competing land use
	Continued poverty
	Lack of existing capacity
	(Low) price of carbon
	Population growth
	Transaction costs
	Monitoring costs
Risk-related	Delay on returns/Slow system response/Permanence
	Leakage/fire/natural variation
Political/bureaucratic	Lack of political will
	Slow land-planning bureaucracy
	Accounting rules complex/unclear and loopholes
Logistical	Different or scattered owners/different interests
	Large areas unmanaged
	Inaccessible areas
	Biological unsuitability
Educational/societal	Stakeholder perception
	Traditional sector
	Sector/legislation is new

on highly organic soils. In developed countries, a limitation on the applicability of mitigation measures can be that agriculture is already managed relatively effectively, for example, with respect to fertilization, whilst in other parts of developed countries significant potential for mitigation still exists (Richards et al., 2006).

The characteristics of risk-related barriers are similar in all regions. The delay in returns from investment in mitigation and the possibility of leakage/sink reversal increase the risk to farmers and land managers in all economic regions.

The lack of political will to encourage mitigation is a significant factor in all economic regions. Smith et al. (2007b) showed that most mitigation that currently occurs is a co-benefit of non-climate policy, often via other environmental policies put in place to promote water quality, air quality, soil fertility, conservation benefits, etc. Indeed, Smith et al. (2005b) showed that even in developed countries (the European Union), little of agriculture's mitigation potential is projected to be realized by 2010 due to lack of incentives to encourage mitigation practices. Bureaucracy can be a significant barrier in all regions, but is especially prevalent in areas where land-planning decisions are slow.

Large unmanaged areas and inaccessibility are barriers mainly in developing countries and countries with economies in transition, with most developed countries having a communications and transport infrastructure to minimize this barrier, despite the very large areas covered by some developed countries.

In terms of educational/societal barriers, traditional practice and stakeholder perception continue to be barriers in all economic regions, though the regional characteristics of these barriers vary greatly. Stakeholder perception is also very different in different regions. Barriers concerning the implementation of bioenergy provide a good example of regional differences. Traditional bioenergy in many developing countries is regarded as a "poor person's fuel" which presents a barrier to its use, whilst in many developed countries, there may be some public resistance to dedicated energy crop monocultures due to their perceived aesthetic impact in rural areas. Barriers may present differently in different regions.

In addition to the barriers presented under the general categories earlier, many potential mitigation measures present very specific barriers. For example, irrigation might increase productivity and thus return more carbon to the soil, thereby sequestering soil carbon. However, in arid areas, competition for water may be a significant barrier. Whilst this will affect arid regions of developed countries, it is likely to present the greatest barrier in arid developing countries. There are also trade-offs between measures, such as use of animal manure for energy production (dung burning in the developing world) or use as a soil amendment. Such barriers are also more likely to arise in developing countries. Smith et al. (2007b) reviewed some of these trade-offs in more detail.

Overcoming the barrier of competition with other land uses will necessitate a holistic consideration of mitigation potential for the land-use sector. It is important that forestry and agricultural land management options are considered within the same framework to optimize mitigation solutions. Costs of verification and monitoring could be reduced by clear guidelines on how to measure, report, and verify greenhouse gas emissions from agriculture. Transaction costs, on the other hand, will be more difficult to address. The process of passing the money and obligations back and forth involves substantial transaction costs, which increases with the number of participants. Given the large number of smallholders in many developing countries, the transaction costs are likely to be higher even than in developed countries, where costs can amount to 25% of the market price (Smith et al., 2007b). Organizations such as farmers' collectives, may help to reduce this significant barrier. Farmers in developing countries are in touch with each other, through local magazines or community meetings, providing forums for these groups to set up consortia of interested forefront players. In order for these collectives to work, regimes need to be in place already and it is essential that the credits are actually paid to the local owner.

For a number of practices, especially those involving carbon sequestration, risk-related barriers such as delay on returns and potential for leakage and sink reversal, can be significant barriers. Education, emphasizing the long-term nature of the sink, could help to overcome this barrier, but fiscal policies (guaranteed markets, risk insurance) might also be required.

Education/societal barriers affect many practices in many regions. There is often a societal preference for traditional farming practices and where mitigation measures alter traditional practice radically (not all practices do), education would help to reduce barriers to their implementation.

In summary, the most significant barriers to implementation of mitigation measures in developing countries (and for some economies in transition) are economic, mostly driven by poverty, which is some areas may be exacerbated by a growing population. In developing countries, many farmers/land managers are poor and struggle to make a living from agriculture, with issues of food security and child malnutrition still prevalent in poor countries (Conway & Toenniessen, 1999). Given the challenges many farmers in developing countries are already facing, soil carbon sequestration is a low priority. To begin to overcome these barriers within the agricultural sector, global sharing of innovative technologies for efficient use of land resources and agricultural chemicals, to eliminate poverty and malnutrition, will significantly help in removing barriers that currently prevent implementation of mitigation measures in agriculture (Smith et al., 2007b). Capacity building and education in the use of innovative technologies and best management practices would also serve to reduce the barriers.

More broadly, macroeconomic policies to reduce debt and to alleviate poverty in developing countries, through encouraging sustainable economic growth and sustainable development, would serve to remove barriers to the implementation of climate mitigation measures in agriculture. Farmers can only be expected to consider carbon sequestration when the threat of poverty and hunger are removed. Sequestration measures, however, may also improve food security and profitability (Lal, 2004b) and such measures are more favorable than those which have no economic or agronomic benefit. Such practices are often referred to as "win–win" options, and strategies to implement such measures can be encouraged on a "no regrets" basis (Smith & Powlson, 2003; Smith, 2004; Lal, 2004b).

2.5   Conclusions

Soil carbon sequestration is a process under the control of human management and, as such, the social dimension needs to be considered when implementing soil carbon sequestration practices. Since there will be increasing competition for limited land resources in the coming century, soil carbon sequestration cannot be viewed in isolation from other environmental and social needs. The IPCC (2001) have noted that global, regional, and local environmental issues such as climate change, loss of biodiversity, desertification, stratospheric ozone depletion, regional acid deposition, and local air quality are inextricably linked. Soil carbon sequestration measures clearly belong in this list. The importance of integrated approaches to sustainable environmental management is becoming ever clearer.

In any scenario, there will be winners and losers. The key to increasing soil carbon sequestration, as part of wider programs to enhance sustainability, is to maximize the number of winners and minimize the number of losers. One possibility for improving the social/cultural acceptability of soil carbon sequestration measures, would be to include compensation costs for losers when costing implementation strategies. By far the best option however, is to identify win–win measures, that is, those which increase carbon stocks whilst at the same time improving other aspects of the environment (e.g., improved soil fertility, decreased erosion, or greater profitability), through, for example, improved yield of agricultural or forestry products. There are a number of management practices available that could be implemented to protect and enhance existing carbon sinks now, and in the future, that is, a no-regrets policy. Smith and Powlson (2003) developed these arguments for soil sustainability, but the no-regrets policy option is equally applicable to soil carbon sequestration. Since such practices are consistent with, and may even be encouraged by, many current international agreements and conventions, their rapid adoption should be encouraged as widely as possible.

Carbon sequestration measures should be considered within a broader framework of sustainable development. Policies to encourage sustainable development will make soil carbon sequestration in developing countries more achievable. Current macroeconomic frameworks do not currently support sustainable development policies at the local level. Policies to encourage fair trade, reduced subsidies for agriculture in developed countries, and less onerous interest on loans and foreign debt would encourage sustainable development, which in turn would provide an environment in which carbon sequestration could be considered in developing countries (Trines et al., 2006).

Acknowledgments I would like to thank Gert-Jan Narbuurs and Eveline Trines for interesting discussions regarding barriers to implementations of carbon sequestration and other land-based mitigation measures, which helped greatly in writing Section 2.4.

References

Allen, J. C. (1985). Soil response to forest clearing in the United States and tropics: Geological and biological factors. *Biotropica, 17*, 15–27.

Batjes, N. H. (1996). Total carbon and nitrogen in the soils of the world. *European Journal of Soil Science, 47*, 151–163.

Cannell, M. G. R. (2003). Carbon sequestration and biomass energy offset: Theoretical, potential and achievable capacities globally, in Europe and the UK. *Biomass and Bioenergy, 24*, 97–116.

Cole, V., Cerri, C., Minami, K., Mosier. A., et al. (1996). Agricultural options for mitigation of greenhouse gas emissions. In: R. T. Watson, M. C Zinyowera, R. H Moss & D. J. Dokken (Eds.), *Climate change 1995. Impacts, adaptations and mitigation of climate change: Scientific-technical analyses* (pp. 745–771). New York: Cambridge University Press.

Conway, G., & Toenniessen, G. (1999). Feeding the world in the twenty-first century. *Nature, 402*, C55–C58.

Detwiller, R. P., & Hall, A. S. (1988). Tropical forests and the global carbon cycle. *Science, 239*, 42–47.

Fearnside, P. M. (1997). Greenhouse gases from deforestation in Brazilian Amazonia: Net committed emissions. *Climatic Change, 35*, 321–360.

Freibauer, A., Rounsevell, M., Smith, P., & Verhagen, A. (2004). Carbon sequestration in the agricultural soils of Europe. *Geoderma, 122*, 1–23.

Follett, R. F., Kimble, J. M., & Lal, R. (2000). The potential of U.S. grazing lands to sequester soil carbon. In: R. F. Follett, J. M. Kimble & R. Lal (Eds.), *The potential of U.S. Grazing lands to sequester carbon and mitigate the greenhouse effect* (pp. 401–430). Boca Raton, FL: Lewis Publishers.

Guo, L. B., & Gifford, R. M. (2002). Soil carbon stocks and land-use change: A meta analysis. *Global Change Biology, 8*, 345–360.

IPCC. (1997). IPCC (Revised 1996) *Guidelines for national greenhouse gas inventories: Workbook*. Paris: Intergovernmental Panel on Climate Change.

IPCC. (2000a). *Special report on land use, land-use change, and forestry*. Cambridge: Cambridge University Press.

IPCC. (2000b). *Special report on emissions scenarios*. Cambridge: Cambridge University Press.

IPCC. (2001). *Climate change: The scientific basis*. Cambridge: Cambridge University Press.

Jenkinson, D.S. (1988). Soil organic matter and its dynamics. In A. Wild (Ed.), *Russell's soil conditions and plant growth* (11th ed., pp. 564–607). London: Longman.

Houghton, R. A. (1999). The annual net flux of carbon to the atmosphere from changes in land use: 1850 to 1990. *Tellus, 50B*, 298–313.

Houghton, R. A., Hackler, J. L., & Lawrence, K. T. (1999). The US carbon budget: Contributions from land-use change. *Science, 285*, 574–578.

Lal, R. (1999). Soil management and restoration for C sequestration to mitigate the accelerated greenhouse effect. *Progress in Environmental Science, 1*, 307–326.

Lal, R. (2001). Potential of desertification control to sequester carbon and mitigate the greenhouse effect. *Climate Change, 15*, 35–72.

Lal, R. (2004a). Soil carbon sequestration to mitigate climate change. *Geoderma, 123*, 1–22.

Lal, R., (2004b): Soil carbon sequestration impacts on global climate change and food security. *Science, 304*, 1623–1627.

Lal, R., Kimble, J. M., Follet, R. F., & Cole, C. V. (1998). The potential of U.S. cropland to sequester carbon and mitigate the greenhouse effect. Chelsea, MI: Ann Arbor Press.

Lohila, A., Aurela, M., Tuovinen, J. P., & Laurila, T. (2004). Annual CO_2 exchange of a peat field growing spring barley or perennial forage grass. *Journal of Geophysical Research, 109*, D18116, doi:10.1029/2004JD004715.

Maljanen, M., Martikainen, P. J., Walden, J., & Silvola, J. (2001). CO_2 exchange in an organic field growing barley or grass in eastern Finland. *Global Change Biology, 7*, 679–692.

Maljanen, M., Komulainen, V. M., Hytonen, J., Martikainen, P., & Laine, J. (2004). Carbon dioxide, nitrous oxide and methane dynamics in boreal organic agricultural soils with different soil characteristics. *Soil Biology and Biochemistry, 36,* 1801–1808.

Maltby, E., & Immirzi, C. P. (1993). Carbon dynamics in peatlands and other wetlands soils: Regional and global perspective. *Chemosphere, 27,* 999–1023.

Mann, L. K. (1986). Changes in soil carbon storage after cultivation. *Soil Science, 142,* 279–288.

Metting, F. B., Smith, J. L., & Amthor, J. S. (1999). Science needs and new technology for soil carbon sequestration. In N.J. Rosenberg, R.C. Izaurralde & Malone, E.L. (Eds.), *Carbon sequestration in soils: Science, monitoring and beyond* (pp. 1–34). Columbus, OH: Battelle Press.

Moraes, J. F. L. de, Volkoff, B., Cerri, C. C., & Bernoux, M. (1995). Soil properties under Amazon forest and changes due to pasture installation in Rondônia, Brazil. *Geoderma, 70,* 63–86.

Nabuurs, G. J., Daamen, W. P., Dolman. A. J., Oenema, O., Verkaik, E., Kabat, P., Whitmore, A. P., & Mohren, G. M. J. (1999). *Resolving issues on terrestrial biospheric sinks in the Kyoto Protocol.* Dutch National Programme on Global Air Pollution and Climate Change, Report 410 200 030 (1999).

Neill, C., Melillo, J. M., Steudler, P. A., Cerri, C. C., Moraes, J. F. L. de, Piccolo, M. C., & Brito, M. (1997). Soil carbon and nitrogen stocks following forest clearing for pasture in the Southwestern Brazilian Amazon. *Ecological Applications, 7,* 1216–1225.

Nykänen, H., Alm, J., Lang, K., Silvola, J., & Martikainen, P. J. (1995). Emissions of CH_4, N_2O and CO_2 from a virgin fen and a fen drained for grassland in Finland. *Journal of Biogeography, 22,* 351–357.

Paustian, K., Andrén, O., Janzen, H. H., Lal, R., Smith, P., Tian, G., Tiessen, H., van Noordwijk, M., & Woomer, P. L. (1997). Agricultural soils as a sink to mitigate CO_2 emissions. *Soil Use and Management, 13,* 229–244.

Richards, K. S., Sampson, R. N., & Brown, S. (2006). Agriculture & Forestlands: U.S. Carbon Policy Strategies. Arlington, TX: Pew Center on Global Climate Change. Available at www.pewclimate.org.

Robertson, G. P., Paul, E. A., & Harwood, R. R. (2000). Greenhouse gases in intensive agriculture: Contributions of individual gases to the radiative forcing of the atmosphere. *Science, 289,* 1922–1925.

Schimel, D. S. (1995). Terrestrial ecosystems and the carbon-cycle. *Global Change Biology, 1,* 77–91.

Schimel, D. S., House, J. I., Hibbard, K. A., Bousquet, P., Ciais, P., Peylin, P., et al. Braswell, B. H., Apps, M. J., Baker, D., Bondeau, A., Canadell, J., Churkina, G., Cramer, W., Denning, A. S., Field, C. B., Friedlingstein, P., Goodale, C., Heimann, M., Houghton, R. A., Melillo, J. M., Moore, B., Murdiyarso, D., Noble, I., Pacala, S. W., Prentice, I. C., Raupach, M. R., Rayner, P. J., Scholes, R. J., Steffen, W. L., & Wirth, C. (2001). Recent patterns and mechanisms of carbon exchange by terrestrial ecosystems. *Nature, 414,* 169–172.

Schlesinger, W. H. (1999). Carbon sequestration in soils. *Science, 284,* 2095.

Smith, J. U., Smith, P., Wattenbach, M., Zaehle, S., Hiederer, R., Jones, R. J. A., et al. (2005a). Projected changes in mineral soil carbon of European croplands and grasslands, 1990–2080. *Global Change Biology, 11,* 2141–2152.

Smith, P. (2004). Soils as carbon sinks: The global context. *Soil Use and Management, 20,* 212–218.

Smith, P., & Powlson, D. S. (2003). Sustainability of soil management practices: A global perspective. In L. K. Abbott & D.V Murphy (Eds.), *Soil biological fertility: A key to sustainable land use in agriculture* (pp. 241–254). Dordrecht, The Netherlands: Kluwer Academic Publishers.

Smith, P., Powlson, D. S., & Glendining, M. J. (1996). Establishing a European soil organic matter network (SOMNET). In: D.S. Powlson, P. Smith & J.U Smith (Eds.), *Evaluation of soil organic matter models using existing, long-term datasets, NATO ASI Series I, Vol. 38* (pp. 81–98). Berlin: Springer-Verlag.

Smith, P., Powlson, D. S., Glendining, M. J., & Smith, J. U. (1997). Potential for carbon sequestration in European soils: Preliminary estimates for five scenarios using results from long-term experiments. *Global Change Biology, 3*, 67–79.

Smith, P., Falloon, P., Coleman, K., Smith, J. U., Piccolo, M., Cerri, C. C., Bernoux, M, Jenkinson, D. S., Ingram, J. S. I., Szabó, J., & Pásztor, L. (1999). Modelling soil carbon dynamics in tropical ecosystems. In R. Lal, J. M. Kimble, R. F. Follett & B. A. Stewart (Eds.), *Global climate change and tropical soils: Advances in soil science,* CRC press, Boca Raton, Florida, USA. (pp. 341–364).

Smith, P., Powlson, D. S., Smith, J. U., Falloon, P. D., & Coleman, K. (2000). Meeting Europe's climate change commitments: Quantitative estimates of the potential for carbon mitigation by agriculture. *Global Change Biology, 6*, 525–539.

Smith, P., Falloon, P., Smith, J. U., & Powlson, D. S. (Eds.). (2001a). Soil Organic Matter Network (SOMNET): 2001 model and experimental metadata, GCTE Report 7 (2nd ed.), GCTE Focus 3 Office. Wallingford, Oxon (224pp.).

Smith, P., Goulding, K. W., Smith, K. A., Powlson, D. S., Smith, J. U., Falloon, P., & Coleman, K. (2001b). Enhancing the carbon sink in European agricultural soils: Including trace gas fluxes in estimates of carbon mitigation potential. *Nutrient Cycling in Agroecosystems, 60*, 237–252.

Smith, P., Falloon, P. D., Körschens, M., Shevtsova, L. K., Franko, U., Romanenkov, V., Coleman, K, Rodionova, V, Smith, J. U., & Schramm, G. (2002). EuroSOMNET—A European database of long-term experiments on soil organic matter: The www metadatabase. *Journal of Agricultural Science* (Cambridge), *138*, 123–134.

Smith, P., Andrén, O., Karlsson, T., Perälä, P., Regina, K., Rounsevell, M., & Van Wesemael, B. (2005b). Carbon sequestration potential in European croplands has been overestimated. *Global Change Biology, 11*, 2153–2163.

Smith, P., Martino, D., Cai, Z., Gwary, D., Janzen, H. H., Kumar, P., McCarl, B., Ogle, S., O'Mara, F., Rice, C., Scholes, R. J., Sirotenko, O., Howden, M., McAllister, T., Pan, G., Romanenkov, V., Schneider, U., Towprayoon, S., Wattenbach, M., & Smith, J.U. (2007a) Greenhouse gas mitigation in agriculture. *Philosophical Transactions of the Royal Society, B*, 363. doi: 10.1098/rstb.2007.2184.

Smith, P., Martino, D., Cai, Z., Gwary, D., Janzen, H. H., Kumar, P., McCarl, B., Ogle, S., O'Mara, F., Rice, C., Scholes, R. J., Sirotenko, O., Howden, M., McAllister, T., Pan, G., Romanenkov, V., Schneider, U., & Towprayoon, S. (2007b) Policy and technological constraints to implementation of greenhouse gas mitigation options in agriculture. *Agriculture, Ecosystems & Environment, 118*, 6–28.

Trines, E., Höhne, N., Jung, M., Skutsch, M., Petsonk, A., Silva-Chavez, G., Smith, P., Nabuurs, G.J., Verweij, P., & Schlamadinger, B. (2006). *Integrating agriculture, forestry, and other land use in future climate regimes: Methodological issues and policy options.* A Report for the Netherlands Research Programme on Climate Change (NRP-CC) (188pp).

Veldkamp, E. (1994). Organic carbon turnover in three tropical soils under pasture after deforestation. *Soil Science Society of America Journal, 58*, 175–180.

Vlek, P. L. G., Rodríguez-Kuhl, G., & Sommer, R. (2004). Energy use and CO_2 production in tropical agriculture and means and strategies for reduction or mitigation. *Environment, Development & Sustainability, 6*, 213–233.

Chapter 3
Land-Use Change and Global Food Production[1]

Navin Ramankutty, Jonathan A. Foley, and Nicholas J. Olejniczak

Abstract This study reviews the major changes in global distribution of croplands during the twentieth century, when the cropland base diminished greatly (from ~0.75 ha/person in 1900 to ~0.35 ha/person in 1990). This loss of croplands was not globally uniform: more than half the world's population, living in developing nations, lost nearly two-thirds of their per capita cropland base. The distribution of croplands has become increasingly skewed—in 1990, 80% of the population lived off less than 0.35 ha/person. While agricultural yields have generally increased, they have barely kept pace with population growth in developing nations. Overall, the global food production system is becoming increasingly vulnerable to regional disruptions because of our increasing reliance on expensive technological options to increase agricultural production, or on global food trade.

3.1 Introduction

The global land base is fundamental to our success as a civilization. Croplands are the sites of world food production, savannas and grasslands provide areas for grazing, and forests and woodlands are sources of fuelwood, paper, timber, and pharmaceutical products. Yet we have a poor understanding of the condition of these crucial land resources. In this chapter we review the recent history of human agricultural activities across the globe, focusing on the worldwide spread and intensification of cropland ecosystems that has occurred in the last century.

---

[1] This chapter is a reprint of an earlier publication: Ramankutty, N., Foley, J. A., and Olejniczak N. J. (2002). People on the land: Changes in population and global croplands during the 20th century, *Ambio*, *31(3)*, 251–257. Section 3.4 on "Agricultural Production" has been rewritten, the corresponding numbers in Table 3.1 have been updated, and a new figure (Fig. 3.7) has been added.

A.K. Braimoh and P.L.G. Vlek (eds.), *Land Use and Soil Resources.*
© Springer Science + Business Media B.V. 2008

3.2   Where Does Our Food Come From?

One of the clearest manifestations of human activity within the biosphere has been the conversion of natural landscapes to highly managed ecosystems, such as croplands, pastures, forest plantations, and urban areas. Until recently, however, we have only had rough estimates of the extent of human-dominated landscapes. For example, Turner II et al. (1993) indicated that roughly 1.4–1.5 billion ha (an area nearly the size of South America) is in some form of cultivation, whereas about 7 billion ha (nearly half the global land surface area) is in pasture and grassland. While these types of estimates are useful, they only indicate the extent of land use within gross national units. More spatially explicit accounts of land-use practices, with greater geographic detail, are needed to evaluate the environmental impact.

Ramankutty and Foley (1998) presented a new technique for documenting and monitoring cropland areas around the world. The method reconciles satellite-based land-cover imagery and worldwide agricultural census data using a simple statistical technique. The IGBP-DIS 1-km land-cover data set (Belward & Loveland, 1996) was fused with numerous agricultural census records compiled by international and national organizations to create a geographically explicit global map of current croplands (Fig. 3.1).

A major advantage of the technique is that satellite data provides information on the spatial distribution of croplands, whereas statistical analysis reconciles satellite and census information. The Ramankutty and Foley (1998) data set indicates that roughly 1.8 billion ha (approximately 12% of the global land surface area) was in cultivation in the 1990s. This number is on the high end of the 1.5–1.8 billion ha reported by other sources (Matthews, 1983; Richards, 1990; Klein Goldewijk, 2001; FAO, 2004). It should be noted that several studies have shown that official statistics may be underreporting agricultural land area—by as much as 50% in some regions (US Department of Agriculture, 1991; Frolking et al., 1999; Seto et al., 2000). Nevertheless, if the Ramankutty and Foley (1998) global total cropland area is an overestimate, the spatial distribution of crop cover is reasonable.

We now present a brief discussion of agricultural geography for 16 regions of the world, with information compiled from numerous sources (Richards, 1990; US Department of Agriculture, 1994; Ramankutty and Foley, 1998; Central Intelligence Agency, 1999). The geographic distribution of croplands shows that the major cultivation zones of the world lie in regions with agriculturally productive soils and adequate climate conditions (Ramankutty, 2000). The breadbaskets of the world include the Corn Belt of the USA, the Prairie Provinces of Canada, the wheat–corn belt of Europe, paddy in the Ganges floodplain, the wheat- and rice-growing regions of eastern China, the grain-growing regions of the Pampas in Argentina, and the wheat belts of Australia. Less dense cultivation occurs throughout much of the developed world, whereas large portions of Africa are characterized by low- to moderate-intensity subsistence agriculture. Croplands are largely absent in regions characterized by extremely dry or cold climates. For instance, croplands are virtually absent in the subtropical deserts, high alpine regions, and high-latitude zones.

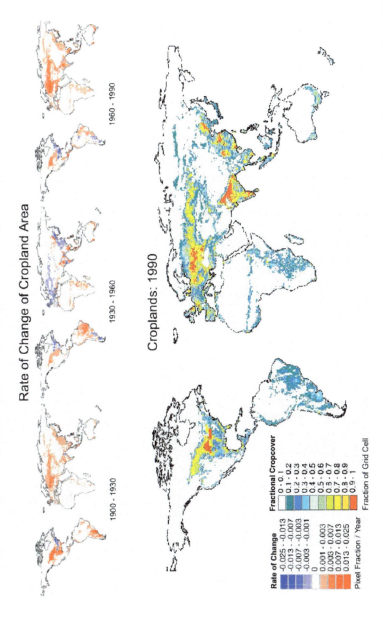

Cropland Intensification and Detensification in the 20th Century

Rate of Change of Cropland Area

1900 - 1930   1930 - 1960   1960 - 1990

Croplands: 1990

Rate of Change
-0.025 - -0.013
-0.013 - -0.007
-0.007 - -0.003
-0.003 - -0.001
0
0.001 - 0.003
0.003 - 0.007
0.007 - 0.013
0.013 - 0.025

Pixel Fraction / Year

Fractional Cropcover
0 - 0.1
0.1 - 0.2
0.2 - 0.3
0.3 - 0.4
0.4 - 0.5
0.5 - 0.6
0.6 - 0.7
0.7 - 0.8
0.8 - 0.9
0.9 - 1

Fraction of Grid Cell

Fig. 3.1 Global cropland change during the twentieth century. (top) cropland rate of change calculated over 30-year intervals from 1900 to 1990. (bottom) global crop cover in 1990. The historical cropland data were derived by statistically synthesizing a satellite-derived land-cover classification data set for 1992 with historical cropland inventory data collected from numerous census organizations

The current extent and intensity of crop cover is a relatively new feature on the Earth's surface. For example, Richards (1990) estimated that there has been greater expansion of cropland areas since World War II than in the eighteenth and early nineteenth centuries combined. Even though many authors have recognized the importance of such large-scale land-use activities, relatively few studies have attempted to quantify the history of human land-use and land-cover change within the biosphere.

Ramankutty and Foley (1999) reconstructed a historical (from 1700 to 1992), geographically explicit data set of cropland areas for the entire globe (Fig. 3.1). This data set was created by statistically combining the 1990s' croplands data set (discussed in Section 3.2) with historical cropland census data (for the last three centuries). Results indicate that croplands expanded by 50% during the twentieth century, from roughly 1.2 million ha in 1900 to 1.8 billion ha in 1990. This net increase in cropland area includes the abandonment of 222 million ha of cropland since 1900.

From this historical data set, we selected data over the 1900–1990 period for the analysis in this study. The general pattern of global crop-cover change shown in this data set (Fig. 3.1) is consistent with the history of human civilizations as well as the patterns of economic development and European settlement (Robertson, 1956; Grigg, 1974; Richards, 1990). The regions of the world with a long history of agricultural practice—Europe, India, China, and Africa—had extensive crop cover in 1900. Cultivation occurred in eastern North America and southeastern Australia following European settlement.

We now summarize three periods of cropland expansion during the twentieth century:

3.2.1   1900–1930

Agriculture in North America migrated westward, with intensification of farming in the Corn Belt in the USA, and cultivation of the Prairie Provinces in Canada. We also see extensive cropland abandonment extending all the way from New England and the Mid-Atlantic states to the Midwestern USA. The Pampas grassland region in Argentina was cleared for crops during this period; parts of Brazil were also beginning to be cultivated. By 1930, cultivation began in southwestern Australia. The rest of the world gradually intensified its cultivation during this time. Some croplands in northeast India were also abandoned.

3.2.2   1930–1960

There was further intensification of crop cover in the American Corn Belt and the Canadian Prairie Provinces. However, cropland abandonment became more extensive in the eastern parts of North America, extending into the southern coastal plains. We see some clearing along the Paranaíba River in Brazil, along with some cropland abandonment to the west. After 1930, crop cover in the Pampas region stabilized and even decreased. The "Virgin Lands" project of the Former Soviet Union [FSU hereafter] led to an intensification of croplands there (although, the data set misrepresents the spatial pattern of this change due to the lack of subnational data; it ought to be an eastward movement of croplands rather than an intensification). While the rate of cropland change increased in the FSU, cropland expansion slowed down in Europe. Cultivation also intensified in India and China, although some cropland abandonment is also seen in India, in the northeast and the southeastern coastal regions. Clearing for croplands intensified in Southeast Asia and Oceania.

3.2.3   1960–1990

The most significant change in croplands was the expansion in southeast Brazil. Cropland abandonment continued in the Pampas. Cropland expansion slowed down in the Midwestern USA, while abandonment in the eastern portions continued. Cropland areas in northern Europe, the FSU, and China stabilized and even decreased in some regions, while it expanded in portions of northeast China. Some croplands were abandoned in Japan. Parts of India continued to clear for crops, while others abandoned croplands. Clearing for cultivation continued in Southeast Asia and Oceania.

We now compare the twentieth-century cropland cover across 16 major regions of the world (Fig. 3.2). As a fraction of total land area, Eastern Europe is the most extensively cultivated region in the world, with more than half its land area in crop cover. However, in absolute terms, the FSU has the largest cropland area. The greatest expansion of croplands in the twentieth century (in absolute amounts) also occurred in the FSU (mainly during expansion into the Virgin Lands) and in northern South America (in the last 50 years, with cultivation in the Brazilian highlands). However, as a percentage of total land area, the greatest cropland expansion occurred in South and Southeast Asia (11% and 18% of their total land area respectively was cleared for cultivation during the twentieth century). In these regions, cropland areas increased exponentially matching an exponentially growing population.

3.3   The Cropland Base

The ability of a region of the world to produce or access food is determined by three factors: (1) availability of an adequate amount of productive cropland, (2) the ability to maintain high crop yields on that land (often with the aid of expensive inputs, such as fertilizers, pesticides, and irrigation), or (3) the ability to purchase and import food from other regions. Other factors such as the economics of food supply within the region and the purchasing ability of individuals and families within each region are also important, but are outside the scope of this chapter.

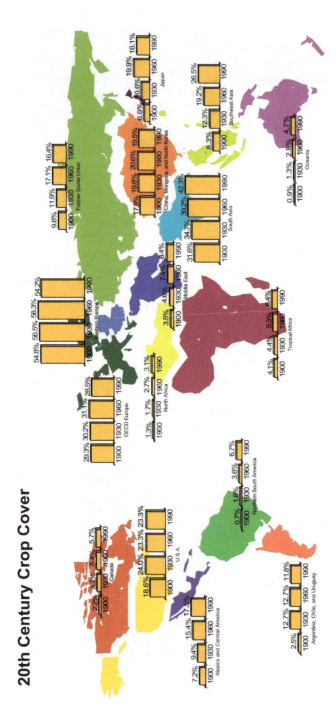

Fig. 3.2 Croplands in the twentieth century. We aggregate the results in Fig. 3.1 into 16 major regions of the world. The results are presented as the fraction of cropland occupying the total land area of each region. As a percentage of total land area, the greatest cropland expansion occurred in South and Southeast Asia, while in absolute amounts, the greatest expansion occurred in the FSU and northern South America

First, we examine the availability of productive cropland around the world. Specifically, we look at the changes in amount of cropland area per person over the twentieth century. Per capita cropland area is an important measure of a region's land base.

3.3.1 Changes in Human Population

During the twentieth century, the world population more than tripled, increasing from approximately 1.5 billion people in 1900 to 5.2 billion in 1990 (Klein Goldewijk, personal communication, 1999). The world's current population of about 6 billion, and is estimated to increase to 8.9 billion by 2050 (United Nations, 2000). In addition to the overall global increases in population, the geographic distribution of human population underwent massive changes during the twentieth century (Fig. 3.3). For example, between 1900 and 1990, the population of northern South America increased by 214 million (681%), compared to the global average population increase of 3,700 million, 236% (Fig. 3.3). Overall, South Asia and China, Mongolia, and North Korea increased in population most rapidly (845 million and 774 million respectively); Canada, Argentina, Chile, and Uruguay had the slowest rates of population growth.

Increasing human population has undoubtedly driven global changes in land cover. However, land-cover change attains global significance through the cumulative addition of locally specific changes (Turner II et al., 1993). Thus, we examine the regional relationships between population change and crop-cover change.

3.3.2 The Changing Distribution of Cropland Resource

Comparing the population and amount of cultivated land in 16 major regions of the world (Fig. 3.4), we find that, in general, regions with higher populations have larger cropland areas. The nature of this relationship has not changed over the twentieth century because it is the greater demand from growing populations that has led to cropland expansion. Furthermore, people tend to live near regions that are agriculturally productive; historically, cities were located close to the major food production centers of the world.

However, the wide scatter around the general linear relationship also reveals an uneven distribution of the cropland base (Fig. 3.4). Developed countries such as the USA and the FSU, with roughly 10–13% of the world population, contain nearly a third of the global cropland area. On the other hand, the populous and poorer nations of the world such as China, Mongolia, and North Korea and South Asia, with roughly 45% of the global population, have only a quarter to a third of the global cropland area.

A good indicator of the global distribution of cropland resources is the per capita cropland area (Fig. 3.5). The twentieth-century changes in per capita cropland area

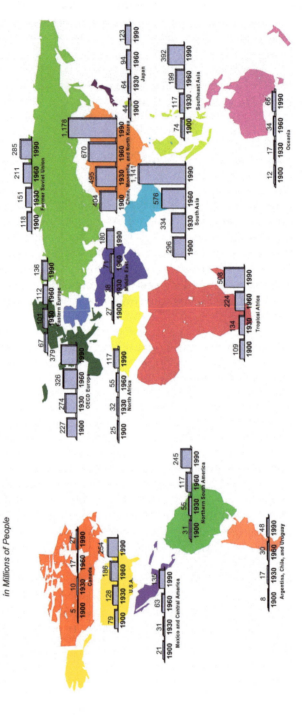

Fig. 3.3 Human population in the twentieth century. Klein Goldewijk and Batjes (1997) compiled historical national- and subnational-level population statistics from various sources. We obtained the data at a country level (Klein Goldewijk, personal communication, 1999) and aggregated to 16 regions. South Asia, China, Mongolia, and North Korea increased in population most rapidly, while Canada, Argentina, Chile, and Uruguay had the slowest rates of population growth. The units are millions of people

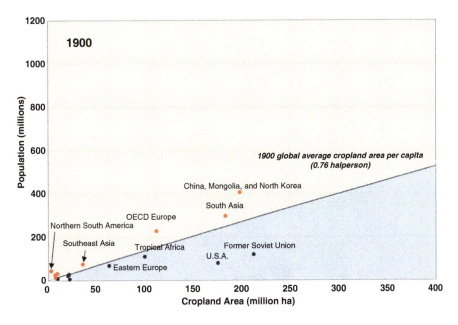

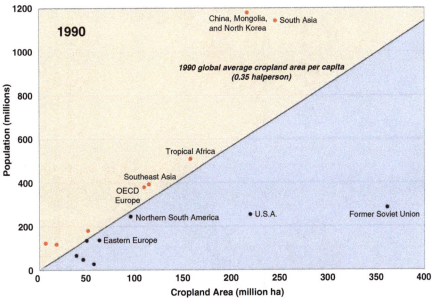

Fig. 3.4 The population of 16 major regions of the world plotted against their cropland area in 1900 and 1990. The dotted line diagonally cutting across the figure represents the global average per capita cropland area in 1900 and 1990. Regions of the world that lie in the blue section of the figure (and represented by blue circles) have relatively higher per capita cropland area, while the regions of the world in pink (and represented by red circles) have lower per capita cropland area with respect to the global average. The global distribution of croplands is skewed: the USA and the FSU, with roughly 10–13% of the world population, have nearly a third of the global cropland area, while China, Mongolia, North Korea, and South Asia, with roughly 45% of the global population, have only a quarter to a third of the global cropland area

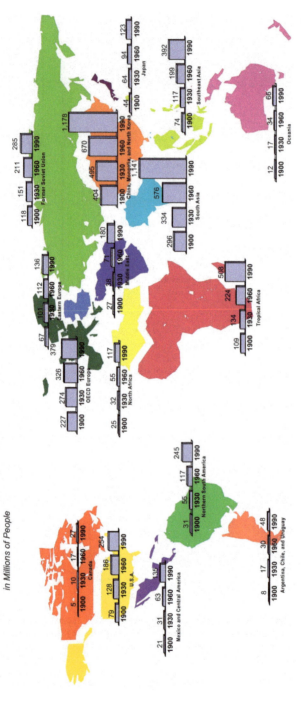

Fig. 3.3 Human population in the twentieth century. Klein Goldewijk and Battjes (1997) compiled historical national- and subnational-level population statistics from various sources. We obtained the data at a country level (Klein Goldewijk, personal communication, 1999) and aggregated to 16 regions. South Asia, China, Mongolia, and North Korea increased in population most rapidly, while Canada, Argentina, Chile, and Uruguay had the slowest rates of population growth. The units are millions of people

in the 16 regions of the world show that population growth was not met by a corresponding increase in cropland expansion both globally and regionally. This is primarily due to two factors. Reason one is that most of the twentieth-century increases in food production were achieved by increased productivity on cultivated land (i.e., by higher yields per unit land area through the use of fertilizers, mechanization, pesticides, and irrigation) Thus, the per capita cropland area decreased in all regions of the world during the twentieth century except in northern South America (Fig. 3.5) (in northern South America, it increased from 1900 to 1930, but decreased afterward, and was still higher in 1990 than in 1900). Globally, the per capita cropland area decreased by more than half, from around 0.75 ha/person in 1900 to only 0.35 ha/person in 1990.

The second reason is that expansion of cropland area did not always occur in the regions with the highest population growth. For example, the FSU had the greatest cropland expansion in the twentieth century (Fig. 3.2), but its population growth was lower than in all the developing nations (Fig. 3.3). On the other hand, South Asia, which had the greatest population increase in the twentieth century, did not have a commensurate expansion of cropland area (Figs. 3.2 and 3.3). Increased food demand in South Asia was met by a large increase in land productivity during the Green Revolution (and besides, fish forms a large portion of the diet). In northern South America, cropland expansion was so large that there was more cropland per person at the end of the century than in 1900. Increased global trade and food aid during the twentieth century undoubtedly provided food to the nations without sufficient cropland base.

It is estimated that a minimum cropland area of 0.5 ha/person is required to provide an adequate diet (Lal, 1989). Of course, this makes certain assumptions about the climatic and soil conditions, and about the level of technology used. However, it is a rough, but useful, number to compare against the cropland base of the different regions of the world during the twentieth century. Canada had the greatest per capita cropland area during the twentieth century (3.4 ha/person, on average), followed by the FSU, Argentina, Chile, and Uruguay, and the USA (roughly 1.5–1.6 ha person). Japan had the lowest per capita cropland base of 0.07 ha person, which is almost 50 times smaller than that of Canada. The greatest loss of per capita cropland was observed in Canada and the USA. In these countries, cropland areas stabilized during the latter half of the twentieth century whereas populations continued to increase. However, these regions already have a large share of the world's cropland base, and relatively fewer people live there.

In relative terms, tropical Africa, South Asia, China, Mongolia, North Korea, Mexico, Central America, the Middle East, and the USA experienced the greatest loss of per capita cropland area during the twentieth century (losing nearly two-thirds between 1900 and 1990). All these regions except for the USA had less than 0.5 ha of cropland per person by 1990. Also, tropical Africa, South Asia, China, Mongolia, and North Korea host more than half of the world's population and are also relatively poor.

The scarcity of the cropland base can be better illustrated by examining the distribution of human numbers versus per capita cropland area (Fig. 3.6). In 1900,

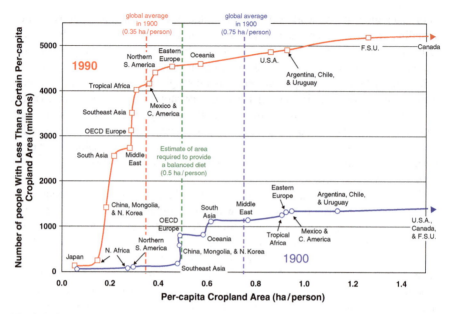

Fig. 3.6 Cumulative distribution of population versus per capita cropland area in 1900 and 1990. The ordinate on the figure indicates the total number of people who live in regions with less than a given per capita cropland area on the abscissa (1900 conditions shown by open circles and 1990 by open squares). Each symbol has been labeled by a region—the values read against the abscissa indicates the region's per capita cropland area, while values on the ordinate indicate the cumulative population of that region and all the regions below it

more than one billion people (74% of the 1900 population) lived below the 1900 global average per capita cropland area of 0.75 ha/person. In 1990, the global average cropland per capita itself decreased to 0.35 ha/person. And in 1990, more than four billion people (nearly 80% of the 1990 population) possessed less than 0.35 ha of cropland per person. Eight regions of the world fall in this category, and five of them—North and tropical Africa, China, Mongolia, North Korea, South Asia, and Southeast Asia—comprise the developing nations of the world. The remaining three regions in this category—Japan, OECD Europe, and the Middle East—are probably wealthy enough to import their food. Furthermore, fish and other seafoods constitute a large proportion of Japanese diets. On the other hand, in 1990, roughly 10% of the global population (living in the developed nations like the USA and the FSU) enjoyed greater than 0.8 ha of cropland per person. In comparison to the estimated minimum cropland area requirement, there were 800 million people in 1900 (52% of the 1900 population) with less than 0.5 ha/person, while in 1990, there were 4.5 billion people (87% of the 1990 population) with less than 0.5 ha/person. Thus, during the twentieth century, the trend indicates that more people (mostly living in developing nations) are living off less land area.

A major caveat regarding the above analysis is that though the level of agro-technology may be essentially similar, the same land area in different parts of the world can produce significantly different amounts of food because of differences in climate and soils. For example, Canada, with a large cropland extent is hampered by a short growing season, whereas the small extent of cropland in Southeast Asia may permit two to three growing seasons. A more complete analysis of regional food status will need to include models of annual crop production. Thus, in Section 3.4, we briefly examine the trends in regional food production over the last decade.

3.4 Agricultural Production

While cropland area is vital for producing food, the agricultural yield of the land (crop production per unit area) is also a critical factor. In fact, much of the increase in food production during the twentieth century came from increases in yield rather than from an expansion of the cropland area. In this section, we briefly examine the status of agricultural yields and total food production during the 1990s.

We compare the population of 16 major regions of the world with their agricultural production in 1990 and 2003 using an estimate of per capita food production (Fig. 3.7; Table 3.1). Regions of the world that lie on the right-hand side of the

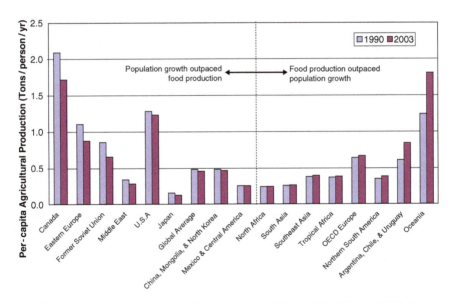

Fig. 3.7 Per capita food production in 16 major regions of the world in 1990 and 2003. Several developing regions of the world have only slightly improved their per capita food production rates between 1990 and 2003. Canada, the USA, Eastern Europe, FSU, the Middle East, and Japan had decreases in per capita food production; however, the former four regions had favorable production rates compared to the global average to begin with, while the latter two regions are wealthy enough to buy food

Table 3.1 Change in per capita agricultural production from 1990 to 2003

	Agricultural production (Million metric tons/year)		Population (Millions of people)		Per capita agricultural production (Tons/person/year)		
	1990	2003	1990	2003	1990	2003	2003–1990
Japan	20	16	124	128	0.16	0.13	−0.03
North Africa	28	37	118	149	0.24	0.25	0.00
Mexico and C. America	35	44	136	171	0.26	0.26	0.00
South Asia	304	400	1158	1486	0.26	0.27	0.01
Middle East	65	71	190	252	0.34	0.28	−0.06
Northern S. America	185	269	500	696	0.37	0.39	0.02
Southeast Asia	87	119	249	306	0.35	0.39	0.04
Tropical Africa	171	217	447	547	0.38	0.40	0.01
Globe	2538	2860	5254	6286	0.48	0.45	−0.03
China, Mongolia, and N. Korea	574	622	1189	1342	0.48	0.46	−0.02
Former Soviet Union	251	190	292	289	0.86	0.66	−0.20
OECD Europe	244	264	380	396	0.64	0.67	0.02
Argentina, Chile, and Uruguay	30	49	49	58	0.61	0.84	0.23
Eastern Europe	132	103	119	117	1.11	0.87	−0.23
USA	327	363	255	294	1.28	1.23	−0.05
Oceania	58	54	28	32	2.10	1.72	−0.37
Canada	25	43	20	24	1.24	1.81	0.57

Note: The agricultural production data (sum of cereals, pulses, and roots, and tubers production) and population data were obtained from the Food and Agricultural Organization's online FAOSTAT database (http://faostat.fao.org). Data were collected at the country level, annually, for the period 1990–2003. The data were aggregated to the 16 regions, and a linear regression was fit to the 14 years of data. The end points of the regression line were chosen to represent the conditions in 1990 and 2003. The data are sorted by the per capita agricultural production rates in 2003. Regions of the world that lie above the global value have relatively lower per capita agricultural production rates in 2003 with respect to the globe, while regions of the world below have higher per capita agricultural production rates.

dashed line had food production outpacing population growth, while those on the left-side had decreases in per capita food production. We also show the global average per capita food production to compare relatively well-off countries to those that are less well-off. We ignore differences in consumption styles and food trade that complicate such a simple analysis.

In 2003, Canada, Oceania, the USA, and Eastern Europe had the highest per capita agricultural production, while Japan, North Africa, Mexico, Central America, and South Asia had the lowest. The direction of change from 1990 to 2003 for the different regions is also seen in Fig. 3.7 and Table 3.1. Global average food production per capita fell slightly from 0.48 tons/ha in 1990 to 0.45 tons/ha in 2003. OECD Europe, Argentina, Chile, Uruguay, and Oceania, all relatively well-off countries in terms of per capita food production compared to the global average, also had food production rates that exceeded their population growth rates between 1990 and 2003. Canada, Eastern Europe, the FSU, and the USA had population growth rates that far exceeded their food production rates; however, they already had favorable production per capita to begin with in the early 1990s. The Middle East and Japan saw decreases in per capita food production between 1990 and 2003, and they were already worse-off than the global average to begin with; however, they are highly dependent on food trade to meet their growing food demands. Per capita food production in China, Mongolia, and North Korea decreased slightly from 1990 to 2003, closely mimicking the global average situation. Several developing regions of the world, all with per capita food production values lower than global average—Mexico and Central America, North Africa, South Asia, Southeast Asia, Tropical Africa, and northern South America—experienced only little improvement in per capita food production values during the 1990–2003 period.

3.5 Future Trends

In the twentieth century, because of technological advances, the food production per capita remained stable and even increased in some places. However, the forecast for the future does not look promising. There will be an additional 2,000 million people on this planet within the next 30 years (United Nations, 2000). Furthermore, food consumption is expected to increase as per capita incomes rise (Daily et al., 1998). In this context, the general trend toward less land per person is disturbing. It implies a greater reliance on continued technological advances in food production; technology that many developing nations cannot afford. Furthermore, there is no fertile frontier remaining to be exploited, since the vast majority of the world's fertile land is already under cultivation. Much of the remaining cultivable land lies in marginal areas or in the richly forested regions of tropical Latin America and Africa (Buringh and Dudal, 1987). Clearing the latter for cultivation implies an enormous loss of valuable forestland and associated biodiversity. Furthermore, we are already losing existing prime farmland to urbanization and soil degradation that will further increase the pressure on the remaining croplands. Thus, our global food production system is becoming increasingly vulnerable because of its sole dependency on technological improvements to meet future food demands; poor nations are likely to be the most affected as they can scarcely afford the expensive technological options or to import food to meet their needs.

3.6 Summary and Conclusion

Humans require a secure and renewable natural resource base to sustain their basic needs and economic activity into the future. However, while deriving natural resources from the terrestrial biosphere, humans also inadvertently modify their environment. The twentieth century has seen a growing human population, increasing agricultural yields, and a resultant decrease in the land resource base. From 1900 to 1990, global population increased 236%, while global cropland area increased 56%, resulting in a halving of the global per capita cropland area.

The loss of land resource base has not been globally uniform. More than half the world's population (living in developing nations constituting South Asia, China, Mongolia, North Korea, and tropical Africa) lost nearly two-thirds of their per capita cropland area during the twentieth century. The impact of rapid population growth in regions with a low cropland base has led to the increasing pressure of more people living off less cropland area per capita. In 1990, nearly 4,000 million people possessed less than the global average per capita cropland area of 0.35 ha/ person, and most of these people lived in poor nations of the world.

A review of food production data indicates that several developing regions of the world have only slightly improved their per capita food production rates between 1990 and 2003. In the future, these nations might have to import much of their food, unless they can continue to enhance the productivity of their land. Canada, the USA, Eastern Europe, FSU, the Middle East, and Japan had decreases in per capita food production; however, the former four regions had favorable production rates to begin with, while the latter two regions are wealthy enough to buy food.

In addition to the loss of per capita cropland area due to the sheer increase in population size, recently, the pressure on the land base has increased further due to the loss of prime farmlands to urbanization and to soil degradation (Kindall and Pimentel, 1994; Gardner, 1996). During the twentieth century, there has been a large shift of humans from rural to urban areas. As most of the urban centers were established close to prime cropland areas, with exploding urban population sizes, human settlements are now encroaching on regions with the best climate and soils for growing crops. For example, in the USA and China, large areas of prime farmland are being lost to urbanization (Mather, 1986; Gardner, 1996). Similarly, with the intensification of cultivation and the expansion of cropland into more marginal areas, large areas of cultivated land are being degraded. A United Nations study (Oldeman et al., 1991) as well as a more recent study by the International Food Policy Research Institute (2000) estimated that roughly 40% of global cropland area is degraded to some degree due to agricultural mismanagement. Thus, in addition to the loss of our per capita cropland base due to the sheer increase in human population, we are also losing prime farmland due to urbanization and soil degradation.

In this study, we have only considered pressures on land due to human population growth. We have to recognize, however, that population growth by itself does not drive changes in land cover. It is also crucial to consider the per capita resource consumption of those individuals. There are vast differences in resource consumption between different economic strata of the human population. To illustrate the importance of per capita resource consumption as an environmental driver, Wackernagel et al. (1997) used the notion of an *environmental footprint* to characterize the land area needed to produce the resources consumed and assimilate the waste generated by a given population. According to Wackernagel et al. (1997), the average American currently requires roughly 20 times as much land as an average Bangladeshi. In fact, if all the people on Earth lived like an average North American, it would require three times the global land area to sustain them. Clearly, differences in economic development, political structure, and culture play a significant role in determining the consumption of natural resources across the globe.

The challenge that lies ahead is to feed an increasing population with increasing per capita consumption. As there is little fertile land remaining to be plowed, future increases in crop production will necessarily have to come from increases in crop productivity. Such increases in productivity will have to be achieved in a sustainable fashion without loss of soil quality and environmental pollution. Can the green revolution that greatly increased productivity in South and Southeast Asia and Latin America also be repeated in tropical South America and Africa? How will such agricultural technology transfers be achieved? Can the failures of the green revolution (salinization, environmental pollution, etc.) be avoided in these new areas of intensified crop production?

References

Belward, A. S., & Loveland, T. R. (1996). The DIS 1 km land cover data set. In The International Geosphere-Biosphere Programme: A study of global change (IGBP) of the International Council of Scientific Unions. *Global change newsletter, 27*, 7–8.

Buringh, P., & Dudal, R. (1987). Agricultural land use in space and time. In M. G. Wolman & F. G. A. Fournier (Eds.), *Land transformation in agriculture, Vol. SCOPE 32* (pp. 3–43). Chichester, UK: Wiley.

Central Intelligence Agency (1999). *The world factbook.* Available at http://www.odci.gov/cia/publications/factbook/index.html.

Daily, G., Dasgupta, P., Bolin, B., et al. (1998). Policy forum: Global food supply—Food production, population growth, and the environment. *Science, 281*, 1291–1292.

FAO (2004). *FAOSTAT data*, Food and Agriculture Organization of the United Nations. Available at http://apps.fao.org.

Frolking, S., Xiao, X. M., Zhuang, Y. H., Salas, W., & Li, C. S. (1999). Agricultural land-use in China: A comparison of area estimates from ground-based census and satellite-borne remote sensing. *Global Ecology and Biogeography, 8*, 407–416.

Gardner, G. (1996). *Shrinking fields: Cropland loss in a world of 8 billion.* Worldwatch Paper 131 (56 pp.). Washington DC: Worldwatch Institute.

Grigg, D. B. (1974). The growth and distribution of the world's arable land 1870–1970. *Geography, 59*, 104–110.

International Food Policy Research Institute (2000). Press release. Available at http://www.cgiar.org/ifpri/pressrel/052500.htm.

Kindall, H. W., & Pimentel, D. (1994). Constraints on the expansion of the global food supply. *Ambio, 23*, 198–205.

Klein Goldewijk, K. (2001). Estimating global land-use change over the past 300 years: The HYDE database. *Global Biogeochemical Cycles, 15*, 417–433.

Klein Goldewijk, C. G. M., & Battjes, J. J. (1997). *A hundred year (1890–1990) database for integrated environmental assessments* (HYDE, version 1.1). Report no. 422514002. Bilthoven, The Netherlands: National Institute of Public Health and the Environment (RIVM).

Lal, R. (1989). Land degradation and its impact on food and other resources. In D. Pimentel & C. W. Hall (Eds.), *Food and natural resources* (pp. 85–140). San Diego, CA: Academic Press.

Mather, A. S. (1986). *Land use* (286 pp.). New York: Longman.

Matthews, E. (1983). Global vegetation and land use: New high resolution data bases for climate studies. *Journal of Climatology and Applied Meteorology, 22*, 474–487.

Oldeman, L. R, Hakkeling, R. T. A., & Sombroek, W. G. (1991). *World map of the status of human induced soil degradation: An explanatory note* (pp. 34). Wageningen, The Netherlands: International Soil Reference and Information Centre.

Ramankutty, N. (2000). The role of croplands in the terrestrial biosphere: Past, present, and future. In *Land resources/institute for environmental studies* (pp. 260). Madison, WI: University of Wisconsin.

Ramankutty, N., & Foley, J. A. (1998). Characterizing patterns of global land use: An analysis of global croplands data. *Global Biogeochemical Cycles, 12*, 667–685.

Ramankutty, N., & Foley, J. A. (1999). Estimating historical changes in global land cover: Croplands from 1700 to 1992. *Global Biogeochemical Cycles, 13*, 997–1027.

Richards, J. F. (1990). Land transformation. In B. L. Turner, W. C. Clark, R. W. Kates, J. F. Richards, J. T. Mathews & W. B. Meyer (Eds.), *The earth as transformed by human action* (pp. 163–178). New York: Cambridge University Press.

Robertson, C. J. (1956). The expansion of the arable area. *The Scottish Geographical Magazine, 72*, 1–20.

Seto, K. C., Kaufmann, R. K., & Woodcock, C. E. (2000). Landsat reveals China's farmland reserves, but they're vanishing fast. *Nature, 406*, 121–121.

Turner II, B. L., Moss, R. H., & Skole, D. L. (1993). *Relating land use and global land-cover change: A proposal for an IGBP-HDP core project*, IGBP report no. 24, HDP report no. 5, International Biosphere-Geosphere Program: A study of global change and the human dimensions of global environmental change programme, Stockholm, (pp. 65).

US Department of Agriculture (1991). *China agriculture and trade report*. Washington, DC: Economic Research Service, U.S. Department of Agriculture.

US Department of Agriculture (1994). *Major world crop areas and climatic profiles* (279 pp.). Agricultural Handbook No. 664, Washington, DC: US Department of Agriculture.

United Nations (2000). *Charting the progress of populations*, United Nations Population Division. Available at http://www.undp.org/popin/wdtrends/chart/contents.htm.

Wackernagel, M., Onisto, L., Linares, A. C., Falfan, I. S. L., Garcia, J. M., Guerrero, A. I. S., & Guerrero, M. G. S. (1997). *Ecological footprints of nations: How much nature do they use? How much nature do they have?* Commissioned by the Earth Council for the Rio + 5 Forum. Toronto: International Council for Local Environmental Initiatives.

Chapter 4
Soil Erosion and Conservation
in Global Agriculture

Hans Hurni, Karl Herweg, Brigitte Portner, and Hanspeter Liniger

Abstract About one-sixth of the world's land area, that is, about one-third of the land used for agriculture, has been affected by soil degradation in the historic past. While most of this damage was caused by water and wind erosion, other forms of soil degradation are induced by biological, chemical, and physical processes. Since the 1950s, pressure on agricultural land has increased considerably owing to population growth and agricultural modernization. Small-scale farming is the largest occupation in the world, involving over 2.5 billion people, over 70% of whom live below the poverty line. Soil erosion, along with other environmental threats, particularly affects these farmers by diminishing yields that are primarily used for subsistence.

Soil and water conservation measures have been developed and applied on many farms. Local and science-based innovations are available for most agroecological conditions and land management and farming types. Principles and measures developed for small-scale as well as modern agricultural systems have begun to show positive impacts in most regions of the world, particularly in wealthier states and modern systems. Much more emphasis still needs to be given to small-scale farming, which requires external support for investment in sustainable land management technologies as an indispensable and integral component of farm activities.

Keywords Soil erosion, soil degradation, small-scale farming, poverty, soil conservation, water conservation, sustainable land management

4.1 Introduction

4.1.1 Background and Research Questions

Global agricultural production basically consists of food for people, feed for livestock, fiber for industry, and fuel for energy. In fact, 95% of the agricultural output is produced on cultivation and grazing lands (Hurni et al., 1996), the rest being the products of marine ecosystems. About 11% of the surface of the world's terrestrial

A.K. Braimoh and P.L.G. Vlek (eds.), *Land Use and Soil Resources.*   41
© Springer Science+Business Media B.V. 2008

ecosystem is used for cultivation, another 25% is used for grazing or grass-cutting, a further 28% is covered by natural and planted forests of different qualities, and about 36% is desert land (FAOSTAT, 2005).

The world's human population has increased by a factor of 2.49 since 1950 (Worldwatch Institute, 2005), causing demand for the above-mentioned products to grow even faster due to increased dietary as well as other per capita demands. Agricultural production, on the positive side, increased even faster, for example, by a factor of 2.75 between 1950 and 1985 for grain cereals (FAOSTAT, 2005), and has remained at about this level. This early increase was due to a number of factors, including advances in agricultural research and technology, plant breeding, increased inputs in minerals and fertilizer, a slight expansion of cultivated land by a factor of 1.13 (FAOSTAT, 2005), and intensification on currently cultivated land.

At present, it appears that both the area of agricultural land and the productivity potentials have reached their upper limits, with little scope to (a) further expand agricultural land sustainably and (b) develop plants that are capable of producing even more per hectare without negatively affecting people and ecology. This, however, is contested by some scientists who claim that there are still vast areas of underutilized arable land, and others who believe that biotechnology will find additional possibilities to enhance plant and animal productivity.

In the last 50 years, pressure on soil and water resources has been considerably accelerated in many places, along with agricultural intensification and particularly expansion. In response, there have been efforts to conserve soil and water and to find means of agricultural production that minimize negative impacts. Much of this reproductive activity, however, is being challenged by persistent poverty, while current actions to reduce poverty in many countries often put even greater strain on agricultural land resources.

Based on these statements the following five research questions were developed as guidelines for the present chapter:

1. How has agricultural development since 1950 affected land resources?
2. How do these changes and prospects relate to processes of soil erosion and land degradation?
3. What is the situation of agriculture today in terms of global poverty, production of food, feed, fiber and fuel, and natural resources?
4. What are the visions for global agriculture in 2050 vis-à-vis human and livestock demands?
5. What potential do soil and water conservation have to achieve sustainable land management now and in future?

It seems obvious that a chapter attempting to answer the above research questions will remain largely interpretative, as only scarce scientific evidence has been produced at the global level that could conclusively answer them. There is, nevertheless, a great need to address these questions, not only in this rather superficial global assessment, but to a much greater extent through intensified scientific research.

4.1.2   Main Hypothesis and Theoretical Concepts

The main hypothesis of this chapter is that global agriculture has the potential to further increase its overall output and feed a growing world population, basically by (a) sustainably enhancing the productivity of small-scale farms and investing in their resource-conserving agricultural technologies and (b) sustainably maintaining industrialized agriculture, which should be capable of feeding growing urban populations.

Research questions 1, 2, 3, and 5 will basically be addressed by synoptically reviewing statistical data and literature, and by interpreting global overviews with maps and country-level indicators. Research question 4 will follow the conceptual framework developed by the International Assessment of Agricultural Science and Technology for Development (IAASTD) as presented in Fig. 4.1.

The implicit theoretical foundation underlying Fig. 4.1 combines a systemic view of agriculture with a normative view of sustainable development. The conceptual framework of IAASTD (2005) in Fig. 4.1 basically looks at agricultural outputs and services, which are determined or influenced by direct and indirect drivers. Indirect drivers are frame conditions such as policies, economic factors, or science and technology, which together influence the direct drivers. The latter act as factors "on the ground" in relation to the production of agricultural goods (i.e., outputs) and services, including forestry and fishery. Direct drivers are thus natural resources, agricultural technology, energy, and

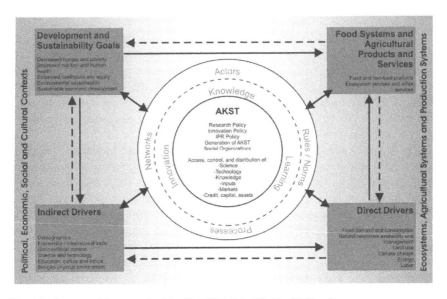

Fig. 4.1 Conceptual framework of the IAASTD. (Modified by H. Hurni)

farm labor. All together the drivers and agricultural outputs and services are oriented toward goals, and can be influenced by these, namely development goals as pursued by individuals, households, communities, and national or international bodies, or sustainability goals now being increasingly introduced at all levels, particularly in view of the overall goal of safeguarding future demands on agriculture without compromising present demands.

Sustainable development is used as a theoretical concept according to the definition by the World Commission on Environment and Development, namely to "satisfy the needs of the present without compromising the needs of future generations" (WCED, 1987). There is thus a claim to assess the intra-generational dimension looking at social, ecological, and economic sustainability, as well as the intergenerational dimension in relation to future generations in the same dimensions.

It is, however, necessary to enhance the conceptual framework of Fig. 4.1. This is currently being discussed in the IAASTD process, with inputs by the main author, the results of which will be published in 2008. This relates to the introduction of a further analytical level, namely typology and analysis of ecological, agricultural, and production systems, which are the basis of agricultural outputs and services and within which drivers have an impact (see Fig. 4.1). This analytical level will allow assessment of changes in the main types of agricultural systems over time, particularly in terms of their resource base, farming systems, and farm household strategies.

4.1.3   Materials and Methods

The present chapter provides a meta-analysis of select information and knowledge generated by science and science-based assessments since 1950. The analysis is illustrated with empirical evidence produced by some of the authors over the last 20–30 years. It attempts to look into the future (2050) using a conceptual framework. While the chapter does not produce new scientific knowledge in a disciplinary field, it is innovative in its integrative interpretation of available scientific information.

The chapter is based on a review of statistical data and literature on agriculture, natural resources, and related frame conditions for agricultural development, such as rural livelihoods, industrialization, urbanization, and the status of human development. Country-level indicators are used to develop a global overview of disparities in development in the agricultural and other sectors. Other global overviews relate to global assessments and related maps, including a critical analysis of their methodologies. The conceptual framework developed by IAASTD (2005) is used as a guideline and critically assessed. For local case studies reference is made to results published primarily by the authors and the programs in which they work.

4.2 Global Agriculture

4.2.1 Agricultural Development Since 1950

The total agricultural output today is about three times higher than it was in 1950. FAOSTAT (2005) reports a 2.75-fold increase for cereals between 1961 and 2003. Besides total output, per capita output has also increased; while the total area of arable land first grew at an average increase of 1.7% from 1961 until 1985, the average increase declined to 0.3% until today (FAOSTAT, 2005). Current estimates indicate that 10–15 million hectares of land are lost each year through erosion and salinization (Pimentel et al., 1993). Although this is less than 1% of the cultivated land, we are threatened by the prospect that without countermeasures soils will be totally depleted in about 200 years from now.

Due to world population growth, from about 2.5 billion people in 1950 to 6.5 billion today (US Census Bureau, 2006), the per capita share of cultivated land has decreased from about 0.4 hectare per person in 1961 to about 0.23 hectare per person today (Fig. 4.2; also see detailed analysis in Chapter 3).

This seeming contradiction between increased output per capita and diminished land per capita is due to the fact that the part of global agriculture influenced by the "green revolution" increased the average per hectare output from about 1 ton to about 3 tons on average for cereals. At present, however, it appears rather difficult to increase average yields beyond that level (Cassman et al., 2003).

The dominant factors in success were plant and animal breeding, mechanization, irrigation, fertilizers, and plant protection. At the same time, less and less people remained in agriculture while this process materialized. In poorer nations, however, where agriculture remained the dominant sector, employing more than 50% of the

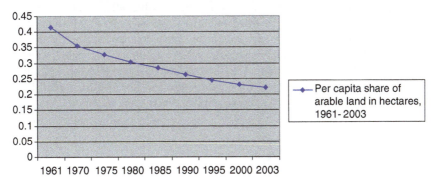

Fig. 4.2 Per capita share of cultivated land in hectares, 1961–2000. (Adapted from FAOSTAT 2005, and US Census Bureau 2006)

population, the takeoff was much slower, and in the poorest countries it did not even take place. A growing disparity has thus evolved since 1950 between industrial agriculture on the one hand, with all its benefits, but also its problems of nonsustainability in energy utilization and biological degradation, and small-scale agriculture on the other hand, where farm families survived basically on a subsistence level, with the associated problems of diminishing farm sizes, overuse of biological resources and human labor, and the resulting nonsustainability here as well.

4.2.2 Agriculture Today

This dichotomy between the two dominant agricultural systems is a fact and will be a challenge in the future. On the one hand, industrial agriculture has been able to feed the growing world population, particularly in urban areas, while on the other hand, small-scale farming in poorer countries has remained stagnant except for diminishing farm sizes due to population pressure in rural areas.

Out of the global population of 6.5 billion people, 1.2 billion live on less than $1 per day, while a further 2.8 billion live on less than $2 per day (Nierenberg and MacDonald, 2004). About 40% of the world's population, or 2.6 billion people, live from agriculture, of which 1.2 billion are actually engaged in agriculture, the rest being children under 15 and old people (MA, 2005). Most of these people are small-scale farmers and therefore part of the global poor mentioned above. About 70% of the world's poorest people living on less than $1 a day are small-scale farmers (MA, 2005). Nutritional security is not automatically guaranteed if a family lives in a rural area and on a farm. On the contrary, many rural people are malnourished, and they are directly vulnerable to the above-mentioned factors of land scarcity and degradation affecting their agricultural production.

It should be noted, however, that soil erosion and conservation are not associated with either type of agriculture (Hurni et al., 1996). Both industrial and small-scale agriculture have problems with natural resource management, such as diminishing biological diversity, soil compaction, soil erosion, salinization, and overfertilization. However, the causes are rather different, a fact that is important when designing strategies for sustainable management of natural resources. Improvements in the industrial agricultural sector can be observed in wealthier nations in the form of bio-farming or precision agriculture. In the small-scale agricultural sector, soil and water management, development of infrastructure and horticulture, a shift to off-farm activities, and a move toward regional planning can be observed.

4.2.3 Agriculture in 2050

In 2050, agriculture and fisheries will need to feed approximately 10 billion people if current population trends materialize (adapted from MA, 2005).

Will global agriculture be able to satisfy the nutritional needs of all these people? Will poverty still persist for nearly 1 billion people at that time, or will the Millennium Development Goals have been achieved by then? And what about the other services to be secured by agriculture? Will natural resources be used in a more equitable and sustainable manner? Will biodiversity be safeguarded even in agricultural areas? Will water be managed in a way that irrigation is possible while drinking and industrial water are still available to a predominantly urban population?

The IAASTD is attempting to develop a series of scenarios to better define the role of research and technology in view of the agricultural challenges ahead. In the opinion of the authors of this chapter, a two-pronged strategy may be the most opportune to pursue, namely:

1. Sustainable production in industrialized agriculture, and
2. Sustainable improvement of small-scale farming

In relation to the first strategy, it will be necessary to further develop mechanized agriculture in most countries with a relatively small agricultural sector of less than 20% of the working population engaged in agriculture. This strategy will have to look at sustainable methods of agricultural production, in terms of plant breeding, energy efficiency, and safe animal and plant protection, without harm to human beings and the environment. The vision here is that on about 40% of all cultivated land, such agriculture could produce about 60% of the global agricultural output with high-yielding crops, and feed about 7 billion people worldwide, mostly those living in urban areas. On the remaining area, that is, on about 60% of the cultivated land, the second vision and strategy calls for small-scale farmers to sustainably produce outputs and services with a smaller per hectare production, but still be able to feed about 3 billion people, that is, mostly themselves and other rural people. This difference in productivity is justified because small-scale farmers will not be able to rationally produce similar amounts per hectare with only their basic labor and few inputs from the other sectors.

In a simple global agricultural output supply model, the first strategy would attempt to produce about 6 tons per hectare on large-scale industrialized farms, while the second strategy would attempt to produce about 3 tons per hectare on small-scale farms. Together, these two dominant agricultural systems could produce enough food, feed, fiber, and maybe even biofuel to meet demand in 2050. The advantage of the two-pronged strategy is that small-scale farmers could continue working with labor-intensive methods on small plots, achieve high employment without the need to migrate to towns, high nutritional security due to subsistence orientation at least for staple crops, and the best possible and sustainable use of marginal lands for agriculture.

The greatest advantage of the above-mentioned strategies is that rural exodus would not be as dramatic as might be expected if all agriculture were industrialized, and environmental damage could also be minimized due to sufficient labor available on small farms for resource-conserving measures. This is a vision for global agriculture in 2050 that finally allows for reduction of the global threat of land degradation that exists today.

4.3   Soil Erosion and Global Agriculture

4.3.1   *Land Degradation: Issues and Stakes*

Land resources such as soil, water, and biodiversity build the basis for agricultural outputs and services. The livelihoods of a vast majority of the world's land users, and particularly of small-scale farmers, depend directly on these resources and the quality of their services. The fact that small-scale farmers often do not have the means to better protect their resources should not be interpreted as lack of awareness. Because ecosystem protection is costly and time consuming, it is usually handled in a reactive way, that is, only when signs of degradation become threatening. It is interesting to note that on the one hand, crises or degradation of plant, animal, and water resources usually trigger immediate responses and countermeasures. Deterioration of the soil resource, on the other hand, is often recognized only after it has already affected plants, animals, and water, that is, at an advanced stage of soil degradation. Consequently, in the past decades, intensification and expansion of grassland and cropland has led to a number of soil degradation processes with different severity and area coverage, among which soil erosion is considered one of the most dangerous processes (cf. chapter 3.1 Oldeman, 1990). This argument is supported by Pimentel et al. (1993) who explain that "because of humankind's almost total dependency on the land for food, soil erosion represents a real threat to the security of our food supply."

Although the focus on soil erosion in this chapter is well justified, it needs to be kept in mind that all processes of natural resource degradation are closely interconnected among themselves (see Fig. 4.3). These relations are particularly important when discussing mitigation practices that are not only supposed to be ecologically sound, but at the same time economically viable and socially acceptable. For example, soil conservation always integrates management of other resources, such as water and plants, and, therefore, soil conservation must be an integral part of an overall land management strategy at the household and community level.

The "Global Assessment of Human Induced Soil Degradation" (GLASOD) (Oldeman et al., 1990) has so far been the only attempt to estimate the problem at the global level. In 1990, Oldeman et al. published a world map known as GLASOD, showing the extent of human-induced soil degradation. Following consultation with experts, GLASOD claimed that 1,964 million hectares, that is, 15.1% of the total land surface, or about one-third of the land used for agriculture, was affected by all forms of soil degradation. Of the affected area, 55.6% was reported as damaged by water erosion, 27.9% by wind erosion, 12.2% by chemical, and 4.2% by physical degradation. Because soil erosion is considered the most widespread soil degradation process, it is a suitable example to explain the severity of environmental problems and the difficulty of finding appropriate solutions. The above-mentioned figures represent the cumulative effect of all previous soil degradation damage originally defined as "since 1950," but probably since much earlier. As the major causes at the global level, GLASOD determined deforestation (30%), overgrazing (35%), and agricultural overuse (28%).

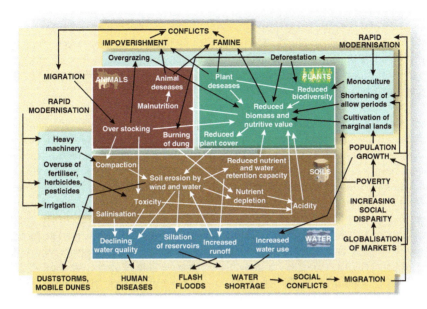

Fig. 4.3 Society–land management–natural resources interactions. (Source: Karl Herwerg)

Table 4.1 Human-induced soil degradation in percent. (Oldeman et al., 1990)

	World	Europe	N & C America	S America	Australasia	Asia	Africa
Category							
Erosion by water	55.6	52.3	67.0	50.6	81.0	58.0	46.0
Erosion by wind	27.9	19.3	25.0	17.2	16.0	30.0	38.0
Chemical deterioration	12.2	11.8	4.0	28.8	1.0	10.0	12.0
Physical deterioration	4.2	16.6	4.0	3.4	2.0	2.0	4.0
Causes							
Deforestation	29.5	38.3	11.3	41.3	12.0	41.0	14.0
Overgrazing	34.5	22.8	24.0	27.8	80.0	26.0	49.0
Overexploitation	6.7	0.2	7.2	4.8	—	6.0	13.0
Agric. activities	28.1	29.3	57.2	26.1	8.0	27.0	24.0
(Bio-)Industrial	1.2	9.4	0.3	—	—	—	—

4.3.2 Current Soil Erosion Rates

In contrast to the so-called developed regions, agricultural land in developing regions has expanded continuously (UNEP, 2003). For example, a detailed study of land-use/land-cover change in the Ethiopian highlands revealed that from 1957 to 1995 cultivated land increased from 39% to 77%, while natural forest declined from 27% to less than 1% in the same period (Gete, 2000). In the observed area soil erosion rates measured on test plots amount to 130 to 170 t/ha/year on cultivated land.

Although such case studies cannot be generalized without an appropriate model, they show the potential hazard of soil degradation related to the shift from forest-land to grazing and cropland, if no suitable countermeasures are taken. Equally, surface runoff during storms increased by a factor of up to 25, with such land-use changes and accelerated land degradation in a long-term empirical study in the Ethiopian highlands (Hurni et al., 2005; see also Chapter 5).

While the GLASOD study presents a cumulative past soil degradation assessment using degree of degradation (light to extreme) and extent of degradation (percentage of the mapping unit affected), there are various estimations of current global erosion rates that present a variety of figures. Several cropland erosion estimates have been compiled by Sundquist (2000). For example, Pimentel (2006) estimates 10 million hectares of cropland to be lost each year due to erosion. Kaiser (2004) quotes an assessment made by the World Soil Information (International Soil Reference and Information Center—ISRIC) in the late 1980s, which found that, "of the 115 million km² of vegetated land on Earth, 17% was degraded, largely through erosion, and 16% could no longer support crops." Other analysts report that the global area of reasonably biologically productive land is on the order of 90 million km² of which about 4.75 million km² is now urbanized (Sundquist, 2007). Pimentel (2003) states that soil erosion rates at the landscape level are highest in Asia, Africa, and South America (30–40 t/ha/year), and lowest in the USA and Europe (17 t/ha/year). Myers (1993) indicates that 75 billion t of fertile topsoil are lost each year from world agricultural systems. Gardner (1996) reports that losses exceeding 100 t/ha/year are common for individual plots, especially on sloping terrain, in many developing countries. Pimentel (1997) estimates topsoil losses in developing countries to average 30 t/ha/year at the landscape level. Also, 80% of the world's agricultural land suffers from moderate to severe erosion, and 10% suffers from slight to moderate erosion (Pimentel et al., 1995). At the same time, Kaiser (2004) quotes a senior scientist at the International Food Policy Research Institute (IFPRI) saying: "As a global problem, soil loss is not likely to be a major constraint to food security." Such general statements, both exaggeration and playing down the erosion problem, should be interpreted with great care.

Apparently, different authors use different sources for their estimates, such as models, plot measurements on farms, sediment data obtained from rivers, etc. Often it remains unclear which factors were included and which were omitted in these statements. Figures measured or estimated with different devices or at different scales are not necessarily comparable. Neither can they be extrapolated without an appropriate model. Consequently, such global estimations and extrapolations of soil loss rates can give only an indication about an average order of magnitude. Proof or disproof is difficult. It is necessary to keep in mind that erosion is a process with great variation both in time and space. Temporal variability of erosion depends greatly on simultaneous occurrence of erosive rainfall and low vegetation cover. Single storms can cause large portions of annual losses and thus lead to great inconsistency in annual values. Similarly, erosion damage is not evenly distributed but rather concentrated on hotspots, which means that, for example, a few locations with extreme damage can alternate with large areas that may barely be affected by erosion (Herweg and Stillhardt, 1999). Also, not all soil that is eroded is "lost" to the ecosystem, but it must be assumed that agricultural production on relatively large areas can be affected by the reduction of topsoil, while only a small area is benefiting from redepositions. Often these depositions are in riverbeds, reservoirs, or waterlogged valley floors and the like, where they are of no direct use for food production. Although erosion may imply a benefit for the recipients of such redeposited topsoil, in the majority of cases it has adverse impacts on agriculture and the environment.

The example of the Ethiopian and Eritrean highlands (Herweg and Stillhardt, 1999; SCRP, 2000a–g), where soil erosion processes were monitored with different devices over a 17-year period, illustrates this variability well. Mean annual erosion measurements on cropland test plots vary from 1 to 212 t/ha/year within and between different regions of the humid highlands. In contrast, corresponding mean annual sediment yield measurements in small catchments with a variety of land-use types show variability from much less than 1 t/ha/year to 25 t/ha/year on average. But even on the same slope, cropland plot values may differ by 40%. The use of models generalizing such findings is not trivial. In the Ethiopian case, Nyssen et al. (2004) compiled the results of several studies that used the adapted Universal Soil Loss Equation (USLE) for soil loss assessment through GIS models. The studies indicated the potential of the model to provide information that is useful in supporting strategic planning of soil and water conservation (SWC). But they also concluded that the model does not provide satisfactory quantitative predictions at the regional/subnational level, which is mainly due to locational inaccuracies for each factor of the model. Given the enormous temporal and spatial variability of soil erosion, and the dependence of measured values on the measurement device with which they have been obtained, the pressing question arises: how reliable and how useful are figures? Neither doomsday scenarios nor neglecting erosion will eventually help achieve more sustainable land management; this can be achieved only by a better differentiation between where and when soil erosion is severe and where and when it is not.

In the next 50 years, the Intergovernmental Panel on Climate Change (IPCC, 2001) expects that climate and environmental changes will most likely become more dramatic, with an increase in nonlinear, that is, potentially catastrophic events. In this case we will be in need of spatially and temporally differentiated information. The Millennium Ecosystem Assessment Synthesis (MA, 2005) assumes that positive scenarios with increased production and simultaneously better ecosystem protection are possible, however, only with significant changes of policies, institutions, and practices, which are currently not yet underway.

4.3.3 Erosion—Productivity Relationships

4.3.3.1 On-Site Effects of Soil Erosion

What erosion really means is understood more easily if its costs, or the (financial) benefits of environmental protection, can be shown. For example, topsoil removal always implies nutrient loss, loss of water by runoff, reduction of rooting depth, and water and nutrient storage capacity, and sooner or later reduced plant production. There are several possibilities to express soil erosion in monetary terms. But again, since the cost of labor, land, and grain differ from place to place, global figures are difficult to obtain. On-site erosion, that is, the erosion damage on the plot or farm where it occurs, can, for example, be expressed by the costs of replacing fertility loss through fertilizers (which does not consider changes in soil organic matter (SOM) and consequently in soil structure), the loss of income due to the reduction in crop yield and biomass production, or price reductions when selling the land. Pimentel et al. (1995) state that in the USA, an estimated 4 billion tons of soil and 130 billion tons of water are lost from the 160 million ha of cropland each year. This translates into an on-site economic loss of more than $27 billion each year, of which $20 billion is for replacement of nutrients and $7 billion for lost water and soil depth. The most significant component of this cost is the loss of soil nutrients. Global costs of soil degradation are estimated to be $400 billion (Pimentel et al., 1995). Despite this, Crosson (1997) concludes that losses due to erosion and other forms of land degradation do not pose a serious threat to the capacity of the global agriculture system to increase yield. He expects a cumulative productivity loss of 17% in 2030, but compared with the expected doubling in global demand for food between 1990 and 2030, he considers the degradation-induced loss to be rather small. The author points out though that there are "hotspots" where soil erosion is more severe and requires immediate attention. In his interpretation, hotspots include selected river basins and are thus related to land degradation on a rather large scale. Again, temporal and spatial variability of soil and productivity loss suggests very careful aggregation of figures. General conclusions at a river basin scale may easily overlook land degradation of "hotspots" at a smaller scale (see Chapter 5), which still remain a threat to millions of small-scale farmers. In addition, global calculations take account of neither the sociocultural value of a soil for a land user nor the ecological value of soils as stores and valves in the ecosystem as a whole.

Stocking (2003) refers to a series of experiments on major tropical soils attempting to determine relationships between crop yield and soil loss, with strong initial yield reduction that declines with time and progressive soil erosion. In all soil scenarios, crop yield becomes almost nil after 1,400 t/ha cumulative soil loss, which reflects a topsoil reduction of only about 140 mm. This figure is surprisingly low and does not correspond with other experiments. For example, Belay (1992) found a very high correlation ($r = 0.96$) between soil productivity and erosion in southern Ethiopia. He estimated that soil productivity declined by 33%, and maize and haricot bean yields by 20–30% in about 50 years, taking into account a measured annual topsoil reduction of 75 t ha year[-1], which would correspond to 3,750 t ha[-1] cumulative soil loss and about 400 mm of topsoil depth. Belay's figure, on the one hand, implies that despite considerable erosion, yield does not quickly drop to zero, but on the other hand, that such a decline in yield by any means has serious effects on the livelihoods of smallholders.

4.3.3.2 Off-Site Effects of Soil Erosion

Aylward (2004) concludes that, regardless of the perceived seriousness of the soil erosion problem, economists and natural scientists agree that downstream or off-site effects of land-use change are in most cases perceived as negative and potentially serious. There is general agreement about the failure of markets to internalize these effects, which are external to land users' decisions. Off-site erosion, that is, the consequences of erosion downslope or downstream, involves costs due to the consequences of sedimentation of dams (reduced production on irrigated land, reduced energy and fish production), sediment load in rivers (reduced water quality and consequent health problems, diminished aesthetic value for recreation areas, impeded navigation and costs of dredging), reclamation activities (sediment removal, buffer strip plantation), and estimating the casualties, costs of injuries, and infrastructure damage due to landslides, mud flows, and flash floods. Clark (1985) estimated the costs of soil erosion to the US water treatment industries to be between $35 million and $661 million per year, while Holmes's (1988) estimates vary between $50 and $500 million. On the other hand, Aylward (2004) presents studies in the Philippines and Panama that computed costs of less than $10/ha/year. Neither argument considers the positive downstream effects of both geological and human-induced soil erosion in terms of renewing soil fertility, as this plays, or played, an important role, for example, along the Nile and Mekong rivers. Aylward (2004) makes clear that generalization of hydrological externalities will not be possible due to the site-specific nature of the biophysical and economic relationships involved. He states that it would be incorrect to assume that hydrological externalities resulting from land-use change are necessarily negative. A priori, the net outcome is indeterminate.

4.4 Soil and Water Conservation in Agriculture

4.4.1 Global Achievements in Soil and Water Conservation

The definition of SWC in a broader perspective encompasses key elements of sustainable land management ranging from addressing degradation to improving soil fertility and productivity (see definition in Box 4.1).

Worldwide, countless efforts have been made to address land degradation, and much experience has been gathered about preventing and mitigating land degradation or rehabilitating badly degraded land. Thus, the wealth of experience in SWC is still scattered and not made easily available to land users and specialists advising land users, planners, and decision-makers as well as researchers. Achievement is slowing due to repetition of the same mistakes instead of building on the local, regional, and international knowledge of available SWC technologies and approaches. While some efforts have been made in documenting degradation, too

> **Box 4.1** Some SWC definitions
>
> Soil and water conservation (SWC) in the context of the WOCAT program is defined as activities at the local level that maintain or enhance the productive capacity of the land in areas affected by, or prone to, degradation. SWC includes prevention or reduction of soil erosion, compaction, and salinity; conservation or drainage of surface and soil water, and maintenance or improvement of soil fertility.
>
> SWC technologies are agronomic, vegetative, structural, and management measures that control land degradation and enhance soil productivity in the field.
>
> SWC approaches are ways and means of support that help to introduce, implement, adapt, and apply SWC technologies on the ground.

few have been made in documenting, monitoring, evaluating, and disseminating experiences with sustainable land management.

Recently, achievements have been made in compiling such knowledge and in making it available in different formats, books, CD ROMs, and on the Internet (WOCAT, 2006). Some results are presented in this chapter, while a comprehensive assessment of efforts and achievements in SWC is in preparation (Liniger and Critchley, 2007).

McNeely and Scherr (2002) describe 36 case studies of sustainable ecosystems based on mainly secondary information from different parts of the world in book form. Bossio et al. (2007) use secondary data and a brief questionnaire to analyze 286 case studies on sustainable agriculture in various regions of the world. The results will be published in database and book form. In 2002, UNEP presents a database and a book on 24 success stories from combating desertification in different areas of the world based on submissions from the field. Two reports focus on Africa: Reij and Steeds (2003) analyze projects and interventions regarding agriculture and rural development in drylands in 15 case studies and Haggblade (2004) describes in a detailed synthesis of 8 case studies on agricultural systems.

4.4.2 Soil and Water Conservation Principles, Measures, and Approaches

4.4.2.1 Prevention, Mitigation, Rehabilitation

Depending on the stage of degradation SWC measures are initiated, we can differentiate between prevention, mitigation, and rehabilitation of land degradation.

Prevention is in the case when land use has maintained the resources and their environmental and productive function and has not yet led to degradation, although the land is prone to degradation. Mitigation is intervention to reduce ongoing degradation and comes at a stage when degradation has already decreased the quality of soil and water resources. The main aim here is to halt further degradation and to start improving (building up) the resources and their functions. Compared to prevention, an already higher investment is needed. However, the impacts of mitigation are visible, which provides a strong motivation for further investments and efforts. Rehabilitation, finally, takes place when the land has already degraded to such an extent that previous land use is no longer possible and land has become practically unproductive.

It needs to be pointed out that efforts and achievements depend very much on the stage of degradation at which they are made. The greatest benefits compared to the investments made are for prevention. This is followed by mitigation, while rehabilitation is the most demanding. Whereas the impacts of rehabilitation efforts can be outstanding, such achievements require critical reflection concerning costs and benefits. A recent analysis showed that major efforts and investments so far have been made to mitigate land degradation and rehabilitate degraded land, whereas attention should move from rehabilitation to prevention (Liniger and Critchley, 2007)

4.4.2.2   SWC Measures in Categories

According to the internationally approved categorization system (Liniger et al., 2002), the measures are grouped first according to land use where the technology is applied, secondly according to the degradation type addressed, and thirdly according to the conservation measures adopted. Differentiation can be made among the following SWC measures.

Agronomic measures (such as mixed cropping, contour cultivation, mulching) are usually associated with annual crops: they are repeated routinely each season or in a rotational sequence, are of short duration and not permanent, do not lead to changes in slope profile, are normally not zoned, and are normally independent of slope. Typical examples are contour planting, direct planting, minimum/noninversion tillage and mulching, and technologies with fertility improvement (e.g., through manuring). Agronomic measures have in recent years received much more attention especially through the Conservation Agriculture (CA) movement, when it became recognized that with little input erosion can be minimized, water used much more efficiently, and soil productivity improved (Fig. 4.4). Such measures have spread in the last decade, both in small-scale and large-scale farming systems.

Vegetative measures (such as grass strips, hedge barriers, and windbreaks) involve the use of perennial grasses, shrubs, or trees, are of long duration, often lead to a change in slope profile, are often zoned on the contour or at right angles to wind direction, and are often spaced according to slope (see Fig. 4.5). Most common and widespread are vegetative strips and cover (mostly grass) and agroforestry systems (intercropping trees with annual and perennial crops, often in a multistory system). These measures are common in humid tropical conditions where often no

Fig. 4.4 Agronomic measure: Conservation Agriculture: zero tillage/direct seeding: Planting with minimum disturbance of the soil, maintaining soil cover and crop rotation, Switzerland

Fig. 4.5 Vegetative measure: Grass strips: Vetiver and Napier grass aligned along contour. After some years the grass strips develop into forward-sloping terraces or even into bench terraces, Tanzania

additional measures are needed due to good ground cover and protection of the vegetation. Under steep slopes, the vegetative measure might be enough (e.g., with tea gardens), while at times structural measures are added. In semiarid conditions, where erosion as well as water stress occurs, vegetative measures also have very positive impacts such as reducing winds (wind erosion) and reducing water loss by evaporation. However, water competition between crops and grasses or trees also occurs. Special management to reduce the competition is needed, for example, the pruning of tree branches and roots. These vegetative measures are often not recognized as SWC measures, especially in traditional land-use systems where erosion has been prevented by them.

Structural measures (such as terraces, banks, bunds, constructions, and palisades) often lead to a change in slope profile; are of long duration or permanent; are carried out primarily to control runoff, wind velocity, and erosion; often require substantial inputs of labor or money when first installed; are often zoned on the contour/against wind direction; are often spaced according to slope; and involve major earth movements and/or construction with wood, stone, concrete, etc. Structural measures are very well recognized and have often been seen as the main measures in combating soil erosion. Recent analysis of SWC efforts made worldwide shows that structural measures are mostly combined with other measures, and that there are many traditional and even ancient terrace systems where maintenance and rehabilitation are needed (Fig. 4.6).

Fig. 4.6 Structural measure: Loess terraces in China

Management measures (such as land-use change, area closure, and rotational grazing) involve a fundamental change in land use, involve no agronomic and structural measures, often result in improved vegetative cover, and often reduce the intensity of use. These measures often apply to grazing land management, where uncontrolled management has led to degradation, and where all other measures do not work without a major change in land management.

Combinations: The measures described above are often combined where they are complementary and thus enhance each other, for example, structural (terrace) with vegetative (grass and trees), with agronomic (contour ploughing; Fig. 4.7).

Technologies can also be arranged in groups, which are familiar to land management specialists and land users, such as CA, manure/compost, vegetative strips and cover, agroforestry, water harvesting, gully control, terraces, grazing land management, and others. For each of these groups single or combined measures can be assigned. Their functioning, costs, and benefits vary greatly.

Fig. 4.7 Combination: Management measures (change from open grazing to hay making and fruit orchard) combined with vegetative (planting of fruit trees) and structural measures (small terraces) in Tajikistan

4.4.2.3 Approaches for Spreading SWC Technologies

Contrary to the SWC technologies, there is no commonly accepted system for the categorization of approaches. However, an analysis of approaches showed that amongst the most important factors underpinning successful development and dissemination of the technologies were: (a) emphasizing systems that improved farmers' yields, (b) providing security of land tenure and access to land, (c) securing markets for produce, and (d) increasing the level of land-user participation in decision-making. Opportunities for the future include tapping into eco/organic labeling, and looking for opportunities to access carbon sequestration markets.

As for the technology, the approach also has to be adapted according to the local and national setting. Illustrating the direct benefits to land users, minimizing inputs, and optimizing the benefits as well as the use of incentives and subsidies, play an important role in supporting land users in their efforts to make a living from the land and at the same time maintain and enhance the various functions of the land.

It has to be noted that some issues driving "conservation" and "good land management" have little to do with trying to stop soil degradation, but are related to soil productivity or reducing the inputs (e.g., the change from conventional tillage to direct seeding) or preferences of land users to do something because they like it more (e.g., planting trees) even though it is not economically beneficial to them (at that time). In some of these activities and changes in the management of land, conservation "comes through the back door."

As the human and natural environments are constantly changing, both the technology and the approach to SWC have to evolve continuously. Thus, it is not surprising that both technologies and approaches in SWC need to be flexible and change over time, and that some of the documented success stories have evolved over a longer period of time and have involved several changes. This can cover rather opposite approaches from regulatory top-down to participatory bottom-up.

A recent assessment of SWC achievements (Liniger and Critchley, 2007) clearly shows that there are many examples of SWC achievements on cropland. However, there are very few examples on grazing land despite degradation being a major problem (35% of grazing lands are seriously affected by degradation (Oldeman et al., 1990)). It appears that the management of grazing land is very challenging and requires further attention and efforts.

4.4.2.4 On-Site Benefits of Soil and Water Conservation

Functioning of SWC

Analysis of the documented technologies showed that in the survey the most frequently mentioned functions of SWC measures addressing different land degradation were, in order of importance: control of dispersed runoff, increase of infiltration, improvement of ground cover, increase of soil fertility, increase of organic matter, increased water stored in the soil, improved soil structure, and control of concentrated runoff.

Combinations of three or more of these functions are common. This is particularly important if soil erosion is combined with other degradation types such as soil fertility decline (chemical degradation) and soil compaction or crusting, where cover improvement and the other mentioned functions are needed.

For soil erosion by water, which is the most frequent degradation type, the following conservation principles can be differentiated: divert/drain runoff and run-on, impede (slow down runoff), retain (avoid runoff), trap (harvest runoff/run-on). In conclusion, the solution lies in better management of rainwater, infiltration in the soil, and surface runoff.

Soil productivity

Based on an analysis by World Overview of Conservation Technologies and Approaches (WOCAT) (Liniger and Critchley, 2007), yield increase (crops, fodder, wood production) was recorded in almost all cases (39 of the 42 case studies presented). Medium to high increases were reported for crops in 60% of the cases, for fodder in almost half, and for wood in 20% of the cases. This implies that some technologies have resulted in an increase in more than one of the above-mentioned categories. Medium to high soil fertility improvement was mentioned in three-quarters of the cases. Thus improvement of soil fertility and productivity is a major, recognized achievement of SWC. Improvement in soil productivity (crops, fodder, trees, other products) is not only related to fertility management, but also to water saving and improved water use (see indirect benefits mentioned later).

Costs and benefits

The most convincing argument for a land user to invest in SWC is an increase in land productivity and hence economic return. However, compiling relevant and reliable information for a proper cost–benefit analysis presented a major challenge to land users and soil and water conservation specialists. Analysis of documented SWC activities reveals that there are marked differences between the various SWC measures (Liniger and Critchley, 2007). All investments as well as benefits need to be looked at, and should be made available to land users and project implementers in order to make informed decisions on what to select and how to combine different measures in a most appropriate and efficient way.

The different technologies need to be looked at in terms of investment and maintenance costs. Most attractive are SWC measures that require low levels of investment as well as low maintenance costs. If little investment is needed, all land users, whether rich or poor, small-scale subsistence or large-scale commercial, have the choice to implement SWC technology. This, for instance, is the case for a number of agronomic and vegetative measures. However, in the case of high investment costs, there is often a need for external support to land users, either by providing incentives and subsidies, credits, loans, and/or grants. High investments are justified when the land-use system is raised to a much higher production level, or when the environmental functions and services of the land require a major investment. For maintenance costs, the major criterion is whether they are needed only over a short period of time (2–3 years) and whether there are additional returns compared to practices without SWC.

For instance, an analysis of conservation agriculture, as an example of an agronomic measure that is being used more and more in mechanized agriculture, showed that investment in new machinery, although cheaper than the conventional machinery, may mean bottlenecks in implementation (IIRR and ACT, 2005; Liniger and Critchley, 2007). Thus, some land users might wait until they have to replace their old machines and tools. However, a number of solutions are available to ease the change, be it by adapting existing machinery and tools or by using contractors. Because the maintenance costs of conservation agriculture are much lower than the costs of conventional land preparation, the benefits are immediately obvious to land users, either in terms of less inputs for the same return, and in drier areas, even much higher returns due to conservation of water.

Vegetative measures seem to be attractive especially in the tropics and subtropics where multistory cropping is possible and attractive as it provides a greater variety of products, especially in small-scale farming at lower altitudes. Vegetative systems are also attractive for fruit production, both in subsistence farming and commercial farming, if some mechanization is possible.

The analysis of structural measures showed a wide range of investment costs, depending on the materials and machinery used, and mainly on the labor costs. Thus, there is a marked difference between small-scale farming systems in developing countries, where investments are relatively cheaper, and mechanized farming systems in the developed world. Costs for structural measures tend to be higher than for other measures. Structures also often mean a loss of the surface available for production. These disadvantages can be compensated by higher returns or by making land use possible in places where crop production and irrigation would not be possible without terracing.

Management measures can achieve remarkable results with low investments. For instance, badly overgrazed and degraded rangelands can recover rapidly after a reduction of grazing pressure and introduction of a rotational system of grazing and resting. Even expensive fencing can be bypassed by stetting up a "social fence," whereby the community agrees where and for how long to rest a piece of land.

From the lessons learnt, the rule of thumb should be to first to assess whether a management measure is feasible, then to consider agronomic and vegetative measures, and to add structural measures "only" when the others are not sufficient. It needs to be pointed out that in the euphoria of implementing SWC, some structural technologies have been pushed "too hard" where other, cheaper measures would have been more appropriate. Due to generally high investment costs and the danger of aggravating erosion if the structural measures are not properly implemented, it can be concluded that other measures should be tried first, and combined with structural measures only if necessary.

Poverty alleviation

Medium to high increase in farm income generated from improved land use through SWC were listed in 88% of the cases. Two-thirds reported a medium to high increase. For each technology, costs and benefits have been assessed, along with the land users' judgment regarding the benefits of the technology in the short and long term. This allows detailed analysis and interpretation. In terms of poverty alleviation, SWC technologies with low investment and maintenance costs and rapid as well as long-term benefits might help small-scale farmers. Several (mainly agronomic and vegetative) measures fulfill these criteria. For development projects, those technologies with "medium to high" initial investments and low maintenance costs but short-term benefits are also possible options. Through external inputs and investment, improvements in SWC could be made and maintained by the local community without additional external support.

4.4.2.5   Indirect and Off-site Benefits of Soil and Water Conservation

In most cases only the direct on-site benefits such as improved production or reduced erosion rates are considered in discussion of the benefits of SWC. Off-site impacts such as less damage on neighbors' fields, roads, houses, and drainage systems, are difficult to assess and quantify. Thus, there is a need for comprehensive assessment of on- and off-site impacts of soil erosion and land degradation and the benefits of SWC, such as initiated in the EU COST 634 program (COST, 2004). However, there are a number of additional issues of global importance that call for proper assessment and attention, such as the indirect benefits of SWC related to carbon sequestration, biodiversity maintenance, desertification mitigation, water management, etc.

The results presented later are based on a recent analysis of 42 case studies throughout the world (Liniger and Critchley, 2007).

Carbon sink

In the global discussion about climate change, the potential for carbon sequestration in the soil is crucial (Robbins, 2004). More than half of the technologies (22 of 42) led to increased soil organic matter (SOM). Given the vast areas of degraded land and the potential of SWC to increase SOM in the topsoil, SWC can be a substantial and long-lasting carbon sink. But once soils are rehabilitated and have reached their climax in terms of SOM, no more additional carbon can be sequestered. Nevertheless, according to Robbins (2004), a substantial potential can be inferred. A proper assessment of this potential to sequester carbon in degraded soils is thus needed. An increase in SOM has additional benefits, including reduced nutrient loss, improved soil structure and water infiltration, reduced erosion, and as a result, increased yields.

Water savings

In addition, 88% of the SWC technologies studied recorded an increase in soil moisture. In 71% of all cases, improvement was rated as medium to high. A second water-related issue is that in one-third of the cases drainage was improved. The functioning of SWC technologies is related to control of dispersed (in 60% of the cases) and concentrated runoff (40%), increase of infiltration (60%), and as a result, an increase in water stored in soil. One-fifth of the cases were explicitly declared to be water harvesting technologies. Reduced water loss through runoff and increased water infiltration and storage in the soil were strongly perceived as leading to greater water-use efficiency. The potential for reducing soil evaporation, especially in drier environments—where 40–70% of the rainfall can be "lost"—has been described clearly in examples of "CA"—a system that combines the three elements of permanent ground cover, minimum disturbance of the soil surface, and crop rotation.

Effects on Water availability, floods, and sedimentation

The most striking water-related off-site benefits of SWC were that three-quarters of the reported case studies reduced downstream flooding and siltation (Liniger and Critchley, 2007). Around half indicated a high to medium impact. Also, 43% indicated reduced river pollution, and about one-third increased stream flow in the dry season. All this information was derived from SWC specialists working with land users, although the impacts have seldom been measured. An exception is the case of Australia's "Green cane trash blanket," where research is assessing the impacts of SWC on rivers and on the Great Barrier Reef. In the absence of such impact assessments, the question arises whether this high rating of off-site impacts is more wishful thinking than proven. Given the growing water-related problems, all impacts mentioned are important for local as well as global discussions of how land use affects water resources. However, there are also a few off-site disadvantages: there was reduced river flow in four cases, although the impact was assessed as "medium" only in one case, and "low" in the others. The cases referred to a situation where terracing and additional irrigation and water harvesting structure reduced flows (see also Hurni et al., 2005).

Desertification mitigation

Water and soil issues and the discussion about desertification are closely linked. Thus, all of the reported cases address degradation of natural resources. Of those, 45% are situated in semi-humid to arid environments, and contribute toward mitigation of desertification. Improvement of infiltration and reduction of runoff and evaporation are closely related to ground cover improvement, which was indicated in 70% of the cases. Soil cover, be it by vegetation or by dead material (e.g., crop residues, leaves), plays a key role in sustainable land management, especially in the tropics and subtropics (Liniger and Thomas, 1998).

Biodiversity enhancement

Although biodiversity has not been a main focus of the WOCAT program, 60% of the cases reported biodiversity enhancement. On cropland this is mainly related to agro-biodiversity, and on grazing land it is linked with improvements in species composition. Again, ground cover and SOM improvement play a key role in above- and below-ground biodiversity. Through the implementation of SWC, mixed farm- ing systems with agroforestry have increased; these favor biodiversity compared with mono-cropping systems.

In this chapter, only selected issues related to soil and water conservation are presented. A more comprehensive analysis and presentation is provided by WOCAT (Liniger and Critchley, 2007), including issues on acceptance and adop- tion and the importance of capacity building, etc.

4.5   The Important Role of Small-Scale Farmers in SWC

As highlighted in the introduction, small-scale farming has special importance in terms of the number of people involved as well as the area covered by small-scale, subsistence, and semi-commercial farming. This section highlights a few issues related to small-scale farming.

Worldwide, huge investments have been made by small-scale farmers and land- use systems have been adapted to local conditions, preventing, mitigating or rehabilitating land degradation. This is illustrated by the WOCAT analysis and recent documentation of bright spots and successful land management practices already mentioned.

CA, as a relatively recent example of an agronomic measure, is based on the three basic principles of minimum soil disturbance, permanent soil cover, and crop mixtures and rotations. Although made popular through large-scale farming, CA has also seen developments and innovations in small-scale farming. Today, CA has been adapted to a wide range of farming conditions, from small-scale to large- scale commercial farming, from the tropics to the medium latitudes (see Figs. 4.8 and 4.9).

Small-scale farmers have developed a whole range of SWC options requiring low inputs and providing increased returns. However, analysis also illustrated that small-scale/subsistence farmers are making considerable investments in SWC (see Figs. 4.10 and 4.11). This is often done through high labor input, due to the fact that labor is more available than any other investment (implements, machines, money). Investments are made especially in soil fertility improvement through manure/com- post, as well as in the construction and maintenance of structural measures, that is,

Fig. 4.8 Small-scale version of conservation agriculture in Kenya. Direct seeding of maize (after wheat) and fertilizer application is done in one pass by a locally manufactured implement. Weeding needs an extra application with herbicides or manual weeding

Fig. 4.9 Computer- and satellite-driven direct seeder for precision conservation agriculture in Australia, including herbicide application only where green weeds are detected by a sensor, where seeding is done and fertilizers applied according to the satellite information about soil fertility and previous yields

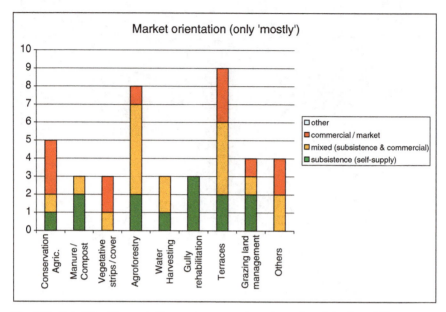

Fig. 4.10 Out of the 42 case studies, we have a good representation of the three different types: Subsistence 31% (13/42), mixed (subsistence and commercial) 40% (17/42), and purely commercial 29% (12/42)

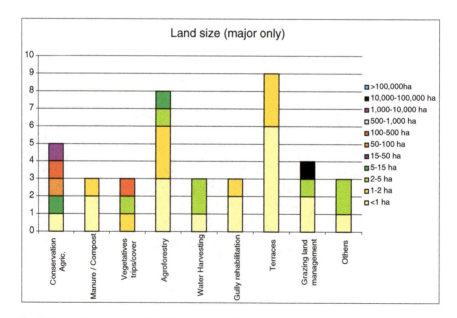

Fig. 4.11 Land-use size for the different case studies

for terraces and water harvesting. The WOCAT analysis showed the striking result that it is mostly small-scale farmers who are making high investments in terraces.

An additional aspect of small-scale farming is its contribution to water conservation and water harvesting, thus making more efficient use of the rainwater. There are many examples of food production being made possible in dry environments without irrigation just by improved water conservation. Given the growing water conflicts caused by irrigation water demand and abstraction, the contribution of improved rainfed agriculture needs to be sufficiently recognized at the local and global levels.

Thus, it is interesting to see that a number of SWC technologies have been developed and work under different farming conditions, following similar principles but using different implements and/or machinery (IIRR and ACT, 2005). An important factor in the successful spread of CA has been the exchange of experiences in different subsistence and large-scale commercial farming communities. Additionally, lots of innovations have been developed by small-scale farmers (Critchley, 1999), preventing degradation, and providing successful examples of land-use (conservation) technologies and approaches which have not been sufficiently recognized. There is a large untapped knowledge pool and farmers are not given credit for their contribution and the positive impact they have on the environment and its functions and services, as well as in the alleviation of poverty. "The many small make a big global contribution."

4.6   Synthesis and Conclusion

In a vision of sustainable agriculture by 2050, the provision of food for 8–12 billion people, animal feed to meet a rapidly growing demand for livestock products, diverse natural fibers for industry, and alternative fuels to substitute for fossil energy have long been the central concerns of strategic planners at the global level. However, the IAASTD, which is currently assessing the contribution of science and knowledge to agricultural development, is adding other services to be provided by agricultural systems for global sustainability, such as ecological services in water provision and regulation, biodiversity conservation, and climate-related functions. Other services include landscape functions for cultural and touristic uses, as well as potential services which are not yet known at present.

Soil degradation, on the other hand, has been widespread on agricultural land, both in modern and small-scale traditional farming systems, affecting not only agricultural outputs but also the above-mentioned additional services. Sustainable land management (SLM), as a consequence, includes the avoidance of soil degradation and the rehabilitation of degraded landscapes, without which it may not be possible to provide the agricultural outputs and services required tomorrow, not to mention in 2050. There exist a variety of technologies to prevent the major threat of soil degradation, and soil erosion by water and wind in different land-use systems worldwide, and many examples have been documented and are ready to be applied. Approaches that support implementation, however, have been less success-fully applied than the technologies; this has resulted in much slower adoption of sustainable soil management on the land than needed.

On the one hand, in wealthier nations, where farmers are only a small fraction of the population today, usually less than 10%, they have been well supported with policies, subsidies, programs, and technologies like conservation agriculture, for which farm implements and supplements are readily made available. Hence processes of soil degradation on their land have been effectively reduced in the last 50–70 years to tolerable levels. Large areas of degradation, nevertheless, still exist in modern states, where in the past less attention was paid to SLM, or fewer resources were made available by comparison with other priorities for develop-ment. Most of these states, however, have taken up the issue particularly in view of the challenges ahead.

On the other hand, small-scale farmers, who constitute the overwhelming major-ity of the world's farming populations today and the largest occupational sector worldwide, have been particularly neglected so far, because most of them live in poorer nations where they constitute between 40% and 90% of the population. Their poverty is caused by limitations in land, labor, and financial capital as well as poor physical and social infrastructure rooted in the weak national economies to which they overwhelmingly contribute. Furthermore, their land is marginal, and of generally poorer quality, although the potential of SLM and increase in productivity are considerable.

As it cannot be expected that the number of small-scale farms, with a current population of nearly 3 billion people, will change significantly in future—even if more and more people move to, and are living in, urban centers—special empha-sis will be needed to support this target group in its efforts to apply soil and water conservation technologies on their farms. This relates particularly to the invest-ment costs that are initially needed to adapt the current farming system to a sys-tem that includes sustainable soil management. Helping small-scale farmers in this risky transition from current practices to a better system will be a task not only for agricultural extension systems in the concerned countries, but for poli-cymakers, who need to adapt policy environments and take investment decisions in state budgets. The international community, finally, has multiple roles to play in sustainable soil management, particularly in support of poorer nations, from information to impact monitoring, to negotiations with countries, and legal support of international actions in science and technology, as well as in policy guidance (Hurni et al., 2006).

Acknowledgments The authors are grateful for the information provided by the following programs and initiatives:

- Centre for Development and Environment, Institute of Geography, University of Berne, partic-ularly through its Soil Conservation Research Program in Ethiopia and its Laikipia Research Program in Kenya
- Swiss National Centre of Competence in Research (NCCR) North-South, University of Berne
- World Overview of Conservation Approaches and Technologies (WOCAT)

- Sustainable Land Management International, a mandate from the Swiss Agency for Development and Cooperation (SDC) to the "International actions for the Sustainable Use of Soils" Working Group of the International Union of Soil Sciences (IUSS)
- International Assessment of Agricultural Science and Technology for Development, IAASTD, World Bank, FAO, and UNEP

References

Aylward, B. (2004). Land-use, hydrological function and economic valuation. In M. Bonell. & L.A Bruijnzeel (Eds.), *Past, present and future hydrological research for integrated land and water management* (pp. 99–120). Cambridge: Cambridge University Press.

Belay, T. (1992). *Erosion: Its effects on properties and productivity of eutric nitosols in Gununo area, southern Ethiopia, and some techniques of its control.* Berne: Geographica Bernensia African Studies Series A9.

Bossio, D. A., Pretty, J., Aloysius N., Noble, A., & Penning de Vries, F. In D. A. Bossio & K. Geheb (Eds.) (2007). *Bright spots. Reversing the trends in land and water degradation: Comprehensive assessment of water management in agriculture.* Oxford: CABI.

Cassman, K. G., Dobermann, A., Walters, D. T., & Yang, H. (2003). Meeting cereal demand while protecting natural resources and improving environmental quality. *Annual Review of Environment and Resources, 28,* 315–358.

Clark E. H. (1985). *Eroding soils: The off-farm impacts.* Washington, DC: The Conservation Foundation.

Critchley W. (Ed.) (1999). Promoting farmer innovation: Harnessing local environmental knowledge in East Africa. Workshop Report No. 2, SIDA Regional Land Management Unit, Nairobi.

Crosson, P. (1997). Will erosion threaten agricultural productivity?. *Environment, 93*(8), 4–31.

COST (European Cooperation in the field of Scientific and Technical Research). (2004). Memorandum of Understanding for the implementation of a European Concerted Research Action designated as COST Action 634 on and off site environmental impacts of runoff and erosion. Brussels: COST Secretariat.

FAOSTAT. (2005). *Food and Agricultural Organization of the United Nations Statistical Databases.* Rome: FAO.

Gardner, G. (1996). Shrinking fields: Cropland loss in a world of eight billion. *World Watch Paper, 131,* 56.

Gete, Z. (2000). *Landscape dynamics and soil erosion process modelling in the north-western Ethiopian highlands. Geographica Bernensis.* Berne: African Studies Series A16.

Haggblade, S. (Ed.) (2004). *Building on successes in African agriculture. 2020 vision for food, agriculture, and the environment.* Focus 12. Washington DC: International Food and Policy Research Institute.

Herweg, K., & Stillhardt, B. (1999). The variability of soil erosion in the highlands of Ethiopia and Eritrea. Average and extreme erosion patterns. Research Report 42. Addis Ababa and Berne: Soil Conservation Research Programme.

Hurni, H., Kebede, T., & Gete, Z. (2005). The implications of changes in population, land use and land management for surface runoff in the upper Nile basin area of Ethiopia. *Mountain Research and Development, 25*(2), 147–154.

Hurni, H., with the assistance of an international group of contributors. (1996). *Precious Earth: From soil and water conservation to sustainable land management.* Wageningen, The Netherlands/Berne: International Soil Conservation Organisation and Centre for Development and Environment.

Hurni, H., Dent, D., Giger, M., & Meyer, K. (Eds.) (2006). Soils on the global agenda. Developing international mechanisms for sustainable land management. Prepared with the support of an international group of specialists of the IASUS Working Group of the International Union of Soil Sciences (IUSS). Berne: Centre for Development and Environment.

Holmes, T. P. (1988). The offsite impact of soil erosion on the water treatment industry. *Land Economics, 64*(4), 356–366.

IAASTD. (2005). International Assessment of Agricultural Science and Technology for Development. An initiative by the World Bank. Available at http://www.agassessment.org.

IPCC (Intergovernmental Panel on Climate Change). (2001). *Climate change 2001: Synthesis report.* A contribution of working groups I, II, and III to the third assessment report of the Intergovernmental Panel on Climate Change [Watson, R.T. and the Core Writing Team (Eds.)]. Cambridge and New York: Cambridge University Press.

IIRR (International Institute of Rural Reconstruction) and ACT (African Conservation Tillage Network). (2005). *Conservation agriculture: A manual for farmers extension workers in Africa.* Nairobi and Harare: International Institute of Rural Reconstruction and African Conservation Tillage Network.

Kaiser, J. (2004). Wounding Earth's fragile skin. *Science, 304,* 1616–1618.

Liniger, H. P., Cahill, D., Thomas, D. B., van Lynden, G. W. J., & Schwilch, G. (2002). *Categorization of SWC technologies and approaches: A global need?* Proceedings of International Soil Conservation Organisation (ISCO) Conference 2002, Vol. III, pp. 6–12, Beijing.

Liniger, H. P., & Critchley, W. (2007). *Where the land is greener. Case studies and analysis of soil and water conservation initiatives worldwide.* Berne: Centre for Development and Environment.

Liniger, H. P., & Thomas, D.B. (1998). GRASS: Ground cover for the restoration of the arid and semi-arid soils. *Advances in GeoEcology, 31,* 1167–1178.

MA (Millennium Ecosystem Assessment). (2005). *Ecosystems and human well-being. Synthesis.* Washington DC: Island Press.

McNeely, J., & Scherr, S.J. (2002). *Ecoagriculture: Strategies to feed the world and save wild biodiversity.* Washington DC: Island Press.

Myers, N. (1993). *Gaia: An atlas of planetary management.* Garden City, NY: Anchor and Doubleday.

Nierenberg, D., & MacDonald, M. (2004). The population story...so far. *World Watch, 17*(5), 14–17.

Nyssen, J., Poesen, J, Moeyersons, J., Deckers, J, Haile, M., & Lang, A. (2004). Human impact on the environment in the Ethiopian highlands: A state of the art. *Earth-Science Reviews, 64,* 273–320.

Oldeman, L. R., Hakkeling, R. T. A., & Sombroek, W. G. (1990). *World map of the status of human-induced soil degradation: An explanatory note.* Wageningen, The Netherlands and Nairobi, Kenya: International Soil Reference and Information Centre and United Nations Environment Programme.

Pimentel, D. (1997). Soil erosion. *Environment, 39*(10), 4.

Pimentel, D. (2006). Soil erosion: A food and environmental threat. *Environment, Development and Sustainability, 8,* 119–137.

Pimentel, D., Allen, J., Beers, A., Guinand, L., Hawkins, A., Linder, R., McLaughlin, P., Meer, B., Musonda, D., Perdue, D., Poisson, S., Salazar, R., Siebert, S., & Stoner, K. (1993). *Soil erosion and agricultural production.* In Pimentel, D. (Ed.), *World soil erosion and conservation* (pp. 277–292). Cambridge: Cambridge University Press.

Pimentel, D., Harvey, C., Resosudarmo, P., Sinclair, K., Kurz, D., McNair, M., Christ, L., Shpritz, L., Fitton, L., Saffouri, R., & Blair, R. (1995). Environmental and economic costs of soil erosion and conservation benefits. *Science, 267,* 1117–1123.

Reij, C., & Steeds, C. (2003). *Success stories in Africa's drylands: Supporting advocates and answering sceptics.* Rome: Global Mechanism of the United Nations Convention to Combat Desertification.

Robbins, M. (2004). *Carbon trading, agriculture and poverty.* Special publication No. 2. Thailand: World Association of Soil and Water Conservation.

SCRP (Soil Conservation Research Programme). (2000a). *Area of Andit Tid, Shewa, Ethiopia: Long-term monitoring of the agricultural environment 1982–1994.* Soil erosion and conservation database. Berne and Addis Ababa, Ethiopia: Centre for Development and Environment, University of Berne and The Ministry of Agriculture, Ethiopia.

SCRP (Soil Conservation Research Programme). (2000b). *Area of Anjeni, Gojam, Ethiopia: Long-term monitoring of the agricultural environment 1984–1994.* Soil erosion and conservation database. Berne and Addis Ababa: Centre for Development and Environment, University of Berne and The Ministry of Agriculture, Ethiopia.

SCRP (Soil Conservation Research Programme). (2000c). *Area of Dizi, Illubabor, Ethiopia: Long-term monitoring of the agricultural environment 1988–1994.* Soil erosion and conservation database. Berne and Addis Ababa: Centre for Development and Environment, University of Berne and The Ministry of Agriculture.

SCRP (Soil Conservation Research Programme). (2000d). *Area of Gununo, Sidamo, Ethiopia: Long-term monitoring of the agricultural environment 1981–1994.* Soil erosion and conservation database. Berne and Addis Ababa: Centre for Development and Environment, University of Berne and The Ministry of Agriculture, Ethiopia.

SCRP (Soil Conservation Research Programme). (2000e). *Area of Hunde Lafto, Harerge, Ethiopia: Long-term monitoring of the agricultural environment 1982–1994.* Soil erosion and conservation database. Berne and Addis Ababa: Centre for Development and Environment, University of Berne and The Ministry of Agriculture, Ethiopia.

SCRP (Soil Conservation Research Programme). (2000f). *Area of Maybar, Wello, Ethiopia: Long-term monitoring of the agricultural environment 1981–1994.* Soil erosion and conservation database. Berne and Addis Ababa: Centre for Development and Environment, University of Berne and The Ministry of Agriculture, Ethiopia.

SCRP (Soil Conservation Research Programme). (2000g). *Concept and methodology: Long-term monitoring of the agricultural environment in six research stations in Ethiopia.* Soil erosion and conservation database. Berne and Addis Ababa: Centre for Development and Environment, University of Berne and The Ministry of Agriculture.

Stocking, M. A. (2003). Tropical soils and food security: The next 50 years. *Science, 302,* 1356–1359.

Sundquist, B. (2007). The earth's carrying capacity, some literature reviews. Topsoil loss – causes, effects, and implications. A global perspective. Internet edition 7, http://home.alltel.net/bsundquist1/.

UNEP (United Nations Environment Programme). (2002). *Success stories in the struggle against desertification.* Nairobi: UNEP.

UNEP (United Nations Environment Programme). (2003). *Global environmental outlook 3.* London: Earthscan.

US Census Bureau. (2006). Total midyear population for the world: 1950–2050. Available at http://www.census.gov/ipc/www/worldpop.html.

WCED (World Commission on Environment and Development). (1987). *Our common future.* Oxford: Oxford University Press.

WOCAT (World Overview of Conservation Approaches and Technologies). (2004). CD-Rom V3: World Overview of Conservation Approaches and Technologies: Introduction – Network – Questionnaires – Databases – Tools – Reports. FAO Land and Water Digital Media Series 9 (rev.). Rome: Food and Agricultural Organization of the United Nations.

WOCAT (World Overview of Conservation Approaches and Technologies). (2006). WOCAT home page. Available at http://www.wocat.org/.

Worldwatch Institute. (2005). *Vital signs 2005. The trends that are shaping our future.* New York and London: W.W. Norton.

Chapter 5
Soil Erosion Studies in Northern Ethiopia

Lulseged Tamene and Paul L.G. Vlek

Abstract Soil erosion is one of the biggest global environmental problems resulting in both on-site and off-site effects. The economic implication of soil erosion is more serious in developing countries because of lack of capacity to cope with it and also to replace lost nutrients. These countries have also high population growth which leads to intensified use of already stressed resources and expansion of production to marginal and fragile lands. Such processes aggravate erosion and productivity declines, resulting in a population–poverty–land degradation cycle.

Rapid population growth, cultivation on steep slopes, clearing of vegetation, and overgrazing are the main factors that accelerate soil erosion in Ethiopia. The annual rate of soil loss in the country is higher than the annual rate of soil formation rate. Annually, Ethiopia losses over 1.5 billion tons of topsoil from the highlands to erosion which could have added about 1.5 million tons of grain to the country's harvest. This indicates that soil erosion is a very serous threat to food security of people and requires urgent management intervention.

To circumvent the impacts of erosion, it is important to know the severity of the problem and the main controlling factors. Since different portions of the landscape vary in sensitivity to erosion due to differences in their geomorphological, geological, and vegetation attributes, it is also necessary to identify high erosion risk areas in order to plan site-specific management interventions. Depending on the prevailing erosion processes and controlling factors, the efficiency of soil conservation measures may vary. This calls for the assessment of the soil conservation potential of different management practices. This study was conducted in northern Ethiopia in order to assess rates of soil loss, investigate controlling factors, and analyze spatial patterns and management alternatives. Section 5.1 reviews the impacts of soil erosion at global and regional scale. Section 5.2 discusses the magnitude of soil erosion in northern Ethiopia based on reservoir survey and Section 5.3 explores its major determinant factors. Section 5.4 applies soil erosion models to identify high erosion risk areas for targeted management intervention and Section 5.5 simulates the potentials of different land management/soil conservation techniques in reducing soil loss of selected catchments. Section 5.6 summarizes the major findings of the study.

A.K. Braimoh and P.L.G. Vlek (eds.), *Land Use and Soil Resources.*   73
© Springer Science + Business Media B.V. 2008

Keywords Soil erosion, soil erosion controlling factors, hot-spot areas of erosion, modeling soil erosion, land use/cover-design, northern Ethiopia

5.1 Soil Erosion and Its Impacts: Global and Regional Perspectives

Soil is being degraded at an unprecedented scale, both in its rate and geographical extent. One of the major causes of soil degradation is soil erosion, which is also among the most serious mechanisms of land degradation and soil fertility decline (Oldeman, 1994). It is the most serious environmental problem affecting the quality of soil, land, and water resources upon which humans depend for their sustenance. El-Swaify (1994) indicated that water erosion accounted for about 55% of the almost 2 billion ha of degraded soils in the world. Lal (1994) compiled worldwide data from different sources and showed that the yield of rainfed agriculture may decrease by about 29% over the next 25 years because of erosion. Pimentel et al. (1995) and Eswaran et al. (2001) estimated world costs of soil erosion to be about US$400 billion per year. A recent study by den Biggelaar et al. (2004) shows that the value of annual production losses for some selected crops worldwide could amount over US$400 million.

Lal (1994) reviewed literature on tropical regions and showed that about 915 million ha of land was degraded due to water erosion. Lal (1995) estimated crop yield reduction of 2–5% per millimeter of soil loss and reported that crop yield reduction due to past erosion in Africa ranges from 2% to 40% with a mean loss of about 8% for the continent. He also showed that soil erosion in Africa has caused yield reductions of about 9%, and if the present trend continues the yield reduction by 2020 may be about 16%. Dregne (1990) identified several regions of Africa where yield reduction due to erosion is as much as 50%.

In Ethiopia, the productivity of the agricultural sector of the economy, which supports about 85% of the workforce, is being seriously affected by soil productivity loss due to erosion and unsustainable land management practices. The average crop yield from a piece of land in Ethiopia is very low according to international standards mainly due to soil fertility decline associated with removal of topsoil by erosion (Sertsu, 2000). Hurni (1990, 1993) estimated that soil loss due to erosion of cultivated fields in Ethiopia amounts to about 42 t ha^{-1} year^{-1}. According to an estimate by FAO (1986), some 50% of the highlands of Ethiopia were already "significantly eroded" in the mid-1980s and erosion was causing a decline in land productivity at the rate of 2.2% per year. The study also predicted that by the year 2010, erosion could reduce per capita incomes of the highland population by about 30%. Taddese (2001) indicated that Ethiopia loses over 1.5 billion tons of soil each year from the highlands by erosion resulting in the reduction of about 1.5 million tons of grain from the country's annual harvest.

Although only limited quantitative data are available on the rates and effects of soil losses, the persistent deterioration of the quality of the cultivated land in

Ethiopia is reflected in degraded slopes, ever-expanding gullies, and associated fragmentation of farm fields. Besides its huge impact on on-site land productivity, soil erosion also causes rapid siltation of streams and reservoirs accelerating storage capacity loss of water harvesting schemes. The rapid water storage capacity losses of dams result in the waste of considerable investments incurred in their construction in addition to failure to achieve food security through surface water harvesting.

Understanding the basic processes and factors that are responsible for soil erosion and associated phenomena is critical to the design and implementation of productive and sustainable agricultural systems. Since all positions of catchments do not experience an equal level of erosion, identification of "hot-spot" (high soil erosion risk) areas is necessary for targeting site-specific management intervention. This study applies an integrated approach to investigate the major geomorphologic and anthropogenic factors controlling water erosion, and seeks to assess the spatial patterns of sediment source areas, and simulates the potentials of different site-specific management interventions at catchment scale.

5.2   Soil Erosion in Northern Ethiopia: Nature and Extent of the Problem

In most of Ethiopia, soil erosion by water is a fundamental problem. A casual visitor is usually amazed by the extent and severity of visible soil erosion of farmlands and grazing areas. Soil erosion is severe in the more barren and mountainous northern highlands. In Tigray (northern Ethiopia), the topsoil, and in some places the subsoil, have been removed from sloping lands, leaving stones or bare rock on the surface (Tilahun et al., 2002). A study by Hurni and Perich (1992) indicated that the Tigray region has lost about 30–50% of its productive capacities compared to its original state some 500 years ago. The high soil-loss rates combined with the prominent gullies in the region comprise one of the most severe land degradation and soil erosion problems in the world (Eweg and Van Lammeren, 1996).

Despite the severity of erosion and its associated consequences in the region, there have been few studies to quantify erosion rates and understand the spatial dynamics of erosion processes at the catchment scale. Some of the studies related to erosion are based on erosion plots (e.g., Nyssen, 2001), which can not be easily extrapolated to basin scale and which are also too few to represent the diverse environments of the region. Furthermore, some of the sediment-yield estimates are based on suspended sediment sampling at gauging stations and may not be reliable, since measurements are not systematic and continuous. The spatial scales of measurements at basin outlets are also generally coarse with limited information to be used for small catchment scale management. There is therefore a need to determine the rate of soil loss and sediment yield at scales that can help narrow the missing link between plot-based and large basin-based studies (Verstraeten and Poesen, 2001). Studies at these scales are important because many of the solutions to

environmental problems such as soil erosion and nonpoint source pollution will
require changes in management at the scale of these landscapes (Wilson and
Gallant, 2000).

Quantitative information on the magnitude and rate of soil loss from catchments
could help to prioritize areas of intervention based on differences in the severity of
erosion problems. This study was conducted in the Tigray region of northern
Ethiopia (Fig. 5.1) to estimate net soil loss/sediment yield in catchments of scale
3–20 km² where different forms of erosion processes and a mosaic of heterogeneous
environmental factors are observed. Reservoir surveys were employed to derive
quantitative estimates of sediment yield from selected catchments. Reservoir
survey methods are more useful and representative because measurements of sediment deposit do not involve generalized statistical models of sediment erosion and
transport or spatial extrapolation of point and plot measurements (Stott et al.,
1988). Data derived from reservoir surveys can also provide a more reliable
indication of sediment loss from catchments than may be obtained from gauging
stations and rating curves (e.g., Walling and Webb, 1981). In order to be able to
extrapolate results to other locations, 11 representative sites were selected on the
basis of size as well as other physical attributes and anthropogenic practices within
catchments.

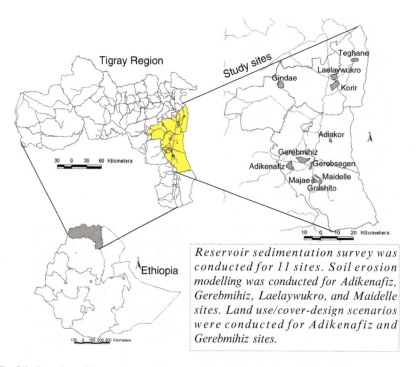

Fig. 5.1 Location of the study area, Tigray, northern Ethiopia. Over 80% of the dams are built in
the food-insecure eastern and southern zones of the region (not fully shown here)

5.2.1 Annual Rates of Soil Losses

The reservoir survey data for the 11 sites shows that net soil losses from catchments ranges from about 3–49 t ha^{-1} year^{-1} with a mean value of 19 t ha^{-1} year^{-1} (Table 5.1).

The relatively high soil-loss variability is due to the wide contrast in environmental variables of the catchments such as terrain, lithology, surface cover, and degree of gully erosion. The magnitude of annual soil loss observed in the study region is generally higher than the "tolerable" soil loss of about 2–18 t ha^{-1} year^{-1} estimated for Ethiopia (Hurni, 1985). All the erosion estimates of the studied sites are above the minimum tolerable limit and five of the catchments have an annual soil loss above the maximum tolerable limit. The soil-loss rates are also beyond the rate of soil formation of 3–7 t ha^{-1} year^{-1} (Hurni, 1983a, b). In addition, the mean net soil loss of 19 t ha^{-1} year^{-1} is high compared with the mean global (15 t ha^{-1} year^{-1}) and mean African (9 t ha^{-1} year^{-1}) sediment-yield rates (Lawrence and Dickinson, 1995).

The observed rates of net soil losses from the studied catchments have direct implications for the storage capacity of water harvesting schemes downstream. In fact, over half of the reservoirs have lost more than 100% of their dead storage capacity (designed to store anticipated sediment until design life) in less than 25% of their expected service time. Most of the reservoirs will be filled with sediment within less than 50% of their projected service lives. Such accelerated loss of storage capacity means that the planned food security improvement scheme, for which the reservoirs were built, is under huge threat. Rapid failure of reservoirs also means a much lower internal rate of return and waste of money spent for the construction of the dams, which otherwise could have been invested for other purposes. The major food security achievement plan of the government is based on surface water harvesting, and this will not be successful unless preventive measures are taken before and after the implementation of the runoff collection schemes.

Knowledge of the rate of soil loss is necessary to identify sensitive areas for intervention based on differences in the severity of soil loss. It is possible to assess

Table 5.1 Annual rate of soil loss from the study catchments in northern Ethiopia

Catchment name	Area (km^2)	Soil loss (t ha^{-1}year^{-1})
Adiakor	2.8	5.0
Adikenafiz	14.0	49.4
Gerebmihiz	19.5	39.2
Gerebsegen	4.0	11.7
Gindae	12.8	19.6
Grashito	5.6	30.2
Korir	18.6	10.6
Laelaywukro	10.0	6.5
Maidelle	8.7	23.6
Majae	2.8	6.8
Teghane	7.0	3.5

which catchments require urgent attention relative to the others and plan appropriate management interventions. In this case, catchments which experience soil loss above a tolerable limit may require urgent management intervention. The next issue is to investigate the major controlling factors in order to be able to design problem-oriented conservation practices.

5.3   Determinants of Soil Erosion in Northern Ethiopia: Causes of Spatial Variability

In order to prescribe appropriate management interventions to tackle soil erosion, knowledge of the factors determining erosion processes is necessary. Quantitative data related to soil loss from basins and corresponding catchment environmental attributes are needed to evaluate cause–effect relationships. These data should be collected from representative locations so that statistical analysis can pinpoint the causes for the observed spatial variability.

Sediment yield from catchments is a reflection of catchment erosion and deposition processes, which are controlled by terrain, soils, surface cover, drainage network, and rainfall-related environmental attributes (Renard and Foster, 1983). Spatial variability in the environmental attributes of catchments may therefore reflect the spatial variability in sediment yield (Verstraeten and Poesen, 2001). Integrated analysis of sediment yield in relation to corresponding environmental attributes of catchments could help identify the dominant factors governing soil-loss variability and evaluate cause–effect relationships at the catchment scale (Dearing and Foster, 1993; Phippen and Wohl, 2003). A combination of bottom-sediment analysis and catchment monitoring also provides a powerful conceptual and methodological framework for improved understanding of drainage basin sediment dynamics (Foster, 1995).

In this study, quantitative data related to catchment environmental attributes were collected based on analysis of digital elevation models, satellite images, and field survey. The catchment attribute data were then integrated with sediment deposition data to evaluate the determinant factors of sediment-yield variability. Statistical and principal component analyses were performed to assess cause–effect relationships.

5.3.1   Controls of Sediment-Yield Variability Among Catchments

Statistical analyses of the relationship between sediment yield from the catchments and corresponding catchment environmental attributes in the study sites show that height (altitude) difference, surface ruggedness (terrain irregularity), surface lithology, degree of gully-related erosion, and surface cover play major roles in soil-loss variability of catchments (Fig. 5.2)

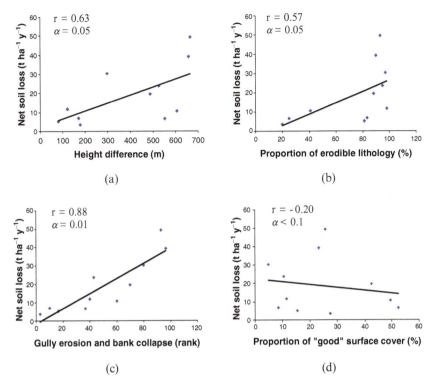

Fig. 5.2 Correlation between net soil loss and selected catchment attributes in Tigray, northern Ethiopia

The height difference is positively and significantly correlated with net soil loss because with increasing altitude difference the runoff and potential energy available to detach and transport soil particles becomes higher. In catchments with high elevation difference, not only is the energy of flow higher, but also the distance to outlets is shorter and the possibility of intermediate deposition is generally lower (Fig. 5.3). The surface ruggedness is also significantly correlated with net soil loss because removal of water and sediment from the channel and watershed surface increases with rapid variation in the slope of catchments.

Since detailed data on soils and their characteristics were not available for all catchments, surface lithology was used as a proxy to assess soil erodibility potential (Fargas et al., 1997). Surface lithology shows a high positive correlation with soil loss; catchments mainly of erodible shale and marl have a higher soil loss than others because such lithologic surfaces are more prone to soil detachment and transport. On the other hand, catchments mainly of less erodible rocks such as sandstone and meta-volcanics show low soil loss partly because of their resistance to erosion.

Figure 5.2c shows that gullies are highly correlated with net soil losses from catchments. This may be due to the fact that a dense network of gullies provides efficient catchment connectivity to deliver sediment to downslope positions

Fig. 5.3 Examples of catchments showing high altitude difference and complex terrain with high potential for sediment transport and delivery

Fig. 5.4 Examples of gully erosion of floodplains and bank collapses in Tigray, northern Ethiopia. The figures were taken in catchments with high sediment yield

(Poesen et al., 2003). Three of the catchments with high soil loss are characterized by spectacular gullies starting from the dams going more than 2–3 km upslope. Gullies are also major sources of sediment in most of the catchments due to bank collapse and through remobilization of sediment deposited in floodplains (Fig. 5.4).

More-gullied catchments produced about double the soil loss of less-gullied catchments. Similar experiences are observed in other regions. For instance, in China, Jiang et al. (1980) reported sediment delivery rates for areas with abundant gullies to be 100–186 t ha^{-1}year^{-1}. Trimble (1974) estimated about 100 t ha^{-1}year^{-1} of soil loss from the southern Piedmont, a major part of which was due to gully erosion. Shibru et al. (2003) showed soil loss of about 25 t ha^{-1} year^{-1} from gullied areas in eastern Ethiopia. Livestock disturbances of gully floors and banks as well as trampling of areas near reservoirs (Fig. 5.3d) worsen the process of gully erosion in most of the study sites. Evidently, attention needs to be given to the rehabilitation/stabilization of gullies and their banks and preventing their destabilization due to livestock trampling.

Gullies may be considered as important factors playing double role: both supply and transport of sediment. This, however, does not exclude the role of sheet/rill erosion in soil loss. In fact, the presence of gullies is a strong indicator that erosion is out of control and the land is entering a critical phase that threatens its productivity (Laflen and Roose, 1998). The presence of eroding and collapsing gullies downslope is mostly associated with a high intensity of runoff and sheet/rill erosion processes within catchments.

Figure 5.2d shows that land use/cover (e.g., proportion of dense bush/shrub) is poorly correlated with soil loss. This is not expected because in theory bush/shrub acts as a buffer and absorbs some of the energy of falling raindrops and running water. The reason for the poor correlation between land use/cover-related factors and soil loss could again be the masking effect of other factors due to high autocorrelation, a common phenomenon of catchment environmental variables (Phippen and Wohl, 2003). Principal component analysis was applied to reduce the dimensionality of the data into a few uncorrelated components and to assess the role of different factors in determining soil loss. After the principal component analysis was run, the effect of surface cover on soil loss became evident as land use/cover types formed the second significant principal component (Tamene et al., 2006).

Another important factor that is expected to play a significant role in increasing runoff and soil loss is slope. Generally, flat land is more stable and soil losses increase rapidly with increasing slope. However, slope is poorly correlated with sediment yield for the catchments studied (not shown). This may be because of the masking effect of some other factors. It is observed that steep-slope areas are correlated with dense surface cover and resistant lithology, while gentle slope areas have generally poor surface cover and erodible lithology (Kirkby et al., 2000). During such circumstances, the separate effects of slope becomes less clear (Rustomji and Prosser, 2001). For instance, the Laelaywukro and Korir sites have complex and steep terrains with a high potential for erosion and sediment yield. But, these catchments are characterized by less erodible lithology and relatively good surface cover, which reduce the severity of erosion and sediment contribution (Fig. 5.5)

In addition, most of the conservation and afforestation efforts which reduce detachment and transport of fine-sediment particles are concentrated on the relatively steep slopes and remote positions. Steep-slope areas are also less prone to

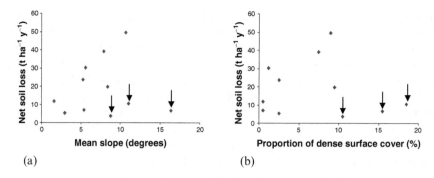

Fig. 5.5 Scatter plot showing the positions of Laelaywukro, Korir, and Teghane catchments when soil loss is plotted against selected attributes. Because of their less erodible lithology and relatively dense surface cover, the net soil loss from the three sites is low despite their relatively steep slope

soil loss because they are mostly located higher uphill having less upslope runoff contributing area and therefore receive less water. In addition, most of the steep slopes are located in the relatively remote parts of the catchments which may have reduced sediment delivery to outlets. The influence of slope is, therefore, neutralized by the combined effects of natural and human factors making its relation to sediment yield unclear.

When the effect of some variable is masked by others, it is necessary to stratify sites, based on some terrain attributes or to exclude the "outliers" from analysis (Lu and Higgitt, 1999). We found the relationship between net soil loss and slope improved when a correlation was performed after excluding the "outlier" catchments from the analysis (Fig. 5.6). This shows the effect of spatial heterogeneity on environmental modeling and demonstrates the necessity for stratification when heterogeneous catchment attributes with complex interactions affect erosion.

Catchment management plays an important role in the severity of erosion. Locations with relatively sound catchment management and conservation practices in place (Laelaywukro and Korir) show rather low soil loss despite the high terrain potential for erosion. However, the widespread conservation efforts undertaken in the region are not always properly maintained and there are several cases where terraces are broken, largely due to livestock trampling or severe runoff. This negates the effectiveness of the conservation measures and even may allow concentrated flow which could enhance erosion and the development of gullies downslope.

A closer look at the environmental factors of the catchments shows that in most instances, erosion and sediment transport-enhancing factors coexist, increasing the potential soil loss and its delivery. For instance, cultivated land is positively correlated with erodible lithology (Fig. 5.7a) and gully erosion is positively correlated with altitude difference (Fig. 5.7b).

Such interactions between the different factors favor both detachment and transport processes, ultimately increasing the potential soil loss from catchments. In general,

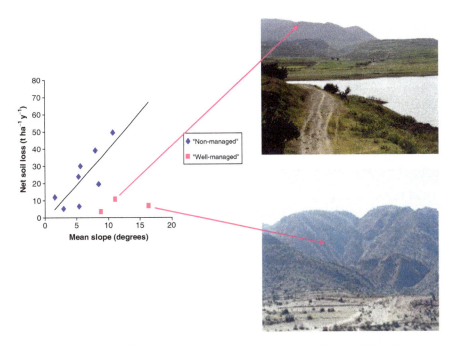

Fig. 5.6 Role of good surface cover and management practices in reducing soil loss from catchments in Tigray, northern Ethiopia. "Nonmanaged" refers to catchment with no enclosures (nonprotected areas) and "well-managed" refers to catchments with protective surface cover such as enclosures and bushes/shrubs

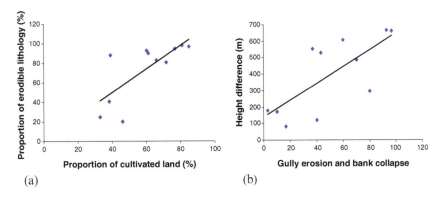

Fig. 5.7 Examples of interactions between different catchment attributes in Tigray, northern Ethiopia

the coexistence of weak, easily detachable lithologic surface, erodible floodplain, and steep terrain (high potential energy to detach, transport, and deliver sediment), as well as poor surface cover (less friction and shear strength of materials), accelerated erosion processes in the study areas. Covering the upland noncultivable areas with suitable vegetation and conservation of gullies could reduce the speed of runoff flow and its erosive capacity, ultimately reducing the rate of erosion and its associated consequences. The question remains, where in the landscape will interventions have greatest effect?

5.3.2   Identifying High Soil Erosion Risk Areas of Catchments: Where Should We Intervene?

As all landscape positions are not equally sensitive to erosion, one important approach to tackling soil loss is to identify where the sources of most of the sediments are within the catchment (Dickinson and Collins, 1998). Given the widely variable rates of soil loss from different landscape units, erosion-control methods to reduce soil loss should focus on the units responsible for the delivery of most sediment. Limited financial resources and land-use restrictions also forbid the application of conservation measures to all areas experiencing soil loss.

Different approaches can be used to identify high erosion risk areas. Soil erosion models represent an efficient means of investigating the physical processes and mechanisms governing erosion rates and identifying high erosion risk areas (Lane et al., 1997; Jetten et al., 2003). A wide range of models are available that differ in their data requirement for model calibration, application, complexity, and processes considered. A good model is one that can satisfy the requirements of reliability, universal applicability, ease of use with minimum data, comprehensiveness in terms of the factors and erosion processes included and the ability to take into account changes in land-use and management practices (Morgan, 1995). However, no single model can satisfy all these requirements or is the "best" for all applications (Grayson and Blöschl, 2000; Merritt et al., 2003). In addition, since most of the existing models are developed in the "data-rich" temperate regions, they can not be directly applied in "data-scarce" tropical montane regions. Generally, models with an adequate physical basis but with optimum data requirement for calibration and validation would be preferable for data-scarce regions.

Recent advances in the development of Geographic Information System (GIS) have promoted the application of distributed soil erosion and sediment delivery models at the catchment scale (Moore et al., 1991; Maidment, 1996; De Roo, 1998). Linking erosion simulation models with GIS provides a powerful tool for land management. It helps to model large catchments with a greater level of detail, allows the presentation of results in more user-friendly formats, and has greater power of data manipulation and the ability to provide a detailed description of catchment morphology through analysis of digital elevation models (De Roo, 1998; Mitas and Mitasova, 1998). GIS technology also allows incorporating physical heterogeneity in catchments and enables basin characteristics controlling sediment detachment, movement, and storage to be considered in a spatially explicit and varying manner (Harden, 1993; Aksoy and Kavvas, 2005).

The Unit Stream Power-based Erosion/Deposition (USPED) model (Mitasova et al., 1996) represents a simple approach to simulating the impacts of complex terrain and various soil- and land-cover changes on the spatial distribution of soil erosion/deposition (Wilson and Lorang, 1999; Wilson and Gallant, 2000). Besides its suitability for complex terrain, its minimum data requirements and its ability to predict the patterns of gullies (Mitasova et al., 2001) has made this model suitable for the Tigray region. As a result, we applied this sort of model in a GIS environment to predict the rate of soil loss and to assess the spatial patterns of erosion/ deposition with the aim of identifying high soil erosion risk areas. The model results were compared with sediment deposition data in reservoirs and data acquired from field surveys.

5.4   Modeling Soil Erosion in the Catchments

5.4.1   Structure of the USPED Model

The USPED model predicts the spatial distribution of erosion and deposition rates for steady-state overland flow with uniform rainfall excess conditions for transport-limited cases of erosion processes (Mitasova et al., 1996). For a transport-limited case of erosion, the model assumes that the sediment flow rate (q_s) is at sediment transport capacity (T) based on Julien and Simons (1985):

$$|q_s| = T = K_t |q|^m [\sin \beta]^n \tag{1}$$

where K_t = transportability coefficient, which is dependent on soil and cover; q = water-flow rate approximated by upslope contributing area; m and n = empirical coefficients that control the relative impact of water and slope terms and reflect different erosion patterns for different types of flow (Mitasova et al., 2001); β = slope angle.

Because there were no experimental data available to develop parameters needed for the USPED, the Universal Soil Loss Equation (USLE) parameters were incorporated to approximate the impacts of soil and vegetation cover and obtain a relative estimate of net erosion and deposition. It was then assumed that sediment flow can be estimated as sediment transport capacity (Mitasova et al., 2001):

$$T = RKCPA_s^m (\sin \beta)^n \tag{2}$$

where R = rainfall erosivity; K = soil erodibility; C = cover and management; P = support practice; A_s = unit contributing area.

Equation (2) defines the availability of stream power for sediment transport and allows the detection of areas with high mass transport capacity. In a three-dimensional

GIS, it is possible to account for other than parallel patterns of sediment flow lines, which may actually be converging or diverging from the given computational cell. This is accomplished by incorporating a topographic parameter describing profile terrain curvature in the direction of the steepest slope and tangential curvature in the direction perpendicular to the profile curvature. The net erosion/deposition is then estimated as the divergence of the pattern T in the computation domain with planar coordinates (x, y) (Mitasova et al., 1996):

$$USPED = div(T.s) = \frac{\partial (T*\cos\alpha)}{\partial x} + \frac{\partial (T*\sin\alpha)}{\partial y} \qquad (3)$$

where s = unit vector in the flow direction; α = aspect of the terrain surface (degrees).

Equation (3) is based on the unit stream power theory and is amenable to landscapes with complex topographies because it explicitly accounts for flow convergence and divergence through the A_s term (Moore et al., 1992). The upslope area is preferred over the slope-length approach at a catchment scale, since upstream area rather than slope-length is the key determinant factor of runoff (Desmet and Govers, 1996; Mitasova et al., 1996).

To predict the rates and spatial patterns of soil erosion, the USLE-based erosion factors (RKCP) adapted for Ethiopian conditions (Hurni, 1985; Machado et al., 1996) were used (Table 5.2)

Once the necessary parameters were determined, the USPED model was applied in four catchments (Adikenafiz, Gerebmihiz, Laelaywukro, and Maidelle) selected for their differences in terrain configuration, land cover and management, intensity of gully erosion, and in the amount of net soil loss based on reservoir surveys.

Table 5.2 KCPR factors (adapted for Ethiopia) and used in the USPED model. (Adapted from Hurni, 1985; Eweg and Lammeren, 1996; and Machado et al., 1996)

Geomorphological unit: (Machado et al., 1996a, b)	K-factor	(Land cover: Hurni, 1985)	C-factor
Erosion remnants with soil cover	0.03	Dense forest	0.001
Erosion remnants without soil cover	0.01	Dense grass	0.01
Badlands	0.04	Degraded grass	0.05
Scarps/denudational rock slopes	0.02	Bush/shrub	0.02
Alluvial fans	0.04	Sorghum, maize	0.10
Alluvial plain and terraces	0.03	Cereals, pulses	0.15
Infilled valleys	0.03	Ethiopian Teff	0.25
(Management type: Hurni, 1985 and Eweg and Lammeren, 1996)	P-factor	Management type	P-factor
Ploughing up and down	1.0	Protected areas	0.50
Strip cultivation	0.80	Ploughing on contour	0.90
Stone cover (40%)	0.80	Terraces	0.60

Rainfall (R) factor (Hurni, 1985) $R = 5.5P - 47$; where P is annual precipitation
Note: Ethiopian Teff—*Eragrostis tef*

5.4.2 Annual Rate of Soil Losses

The predicted mean annual rate of soil loss for each of the four catchments based on the USPED model ranges between 14 t ha^{-1} year^{-1} and 50 t ha^{-1} year^{-1} (Fig. 5.8). The model predicted a higher soil-loss rate than the average soil generation rate of 6 t ha^{-1} year^{-1} (Hurni, 1983a, b) or the maximum tolerable soil-loss rate of 18 t ha^{-1} year^{-1} estimated for the country (Hurni, 1985) for three catchments. About 40% of Adikenafiz and Gerebmihiz, 25% of Maidelle) and 30% of Laelaywukro) are eroding at a rate higher than the maximum tolerable soil-loss rate of 18 t ha^{-1} year^{-1}. More than 35% of each of the three catchments experiences soil-loss rates of greater than 25 t ha^{-1} year^{-1}. An exception is Laelaywukro, where a little less than 20% of the catchment experiences such soil losses. Generally, if the areas experiencing more than 30 t ha^{-1} year^{-1} of soil loss are considered to be in need of conservation, 35% (Adikenafiz) and 45% (Gerebmihiz) catchments fall into this category.

The USPED model predicted high rates of soil loss for the Adikenafiz and Gerebmihiz catchments compared to the others (Fig. 5.8), which may be attributed to their complex terrain and dense network of gullies. The Maidelle catchment has a relatively simple linear slope with an elongated shape, which may not favor erosion. Gullies are not as prominent in Maidelle as they are in the other two sites with serious erosion. In the case of the Laelaywukro catchment, it appears that existing conservation practices and relatively dense bush/shrub cover and the dominance of less-erodible sandstone prevent high net soil loss despite its complex terrain.

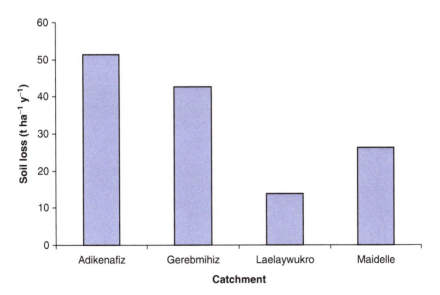

Fig. 5.8 Soil loss predictions of four catchments based on the USPED model in Tigray, northern Ethiopia

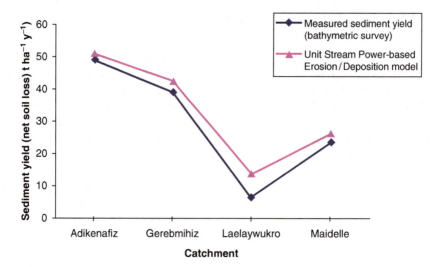

Fig. 5.9 Relationship between reservoir-based sediment-yield estimates and model-based net soil-loss predictions in example catchments of Tigray, northern Ethiopia

Generally, the magnitude of erosion predicted by the USPED model exceeds deposition. For example, for the Adikenafiz catchment, the USPED shows that about 62% and about 27% of the catchment are characterized by erosion and deposition, respectively. This means that soil loss is higher than the amount that can be redistributed within the catchments, increasing the potential for sediment export. The high soil-loss rates predicted along gullies also indicates that the potential for sediment export is higher than for redistribution.

The annual net soil losses from catchments predicted by the model agree very well with annual sediment depositions in reservoirs (Pearson's $r > 0.97$, $\alpha = 0.001$, Fig. 5.9). The root mean square error, which compares the observed (reservoir-based) and predicted (USPED-based) soil-loss estimates for the four sites, also shows an overall error of the USPED model of about $5\,t\ ha^{-1}\ year^{-1}$. This means that the model can be applied in the region to estimate soil loss with an adequate level of accuracy.

5.4.3   Spatial Patterns of Soil Erosion

According to the USPED model, most of the areas experiencing high soil loss are the upper parts of catchments and gullies (Fig. 5.10)

The "wavy" appearance of the erosion maps is due to the fact that the steep sides of gullies have very high erosion rates while the floors of gullies have very high deposition rates.

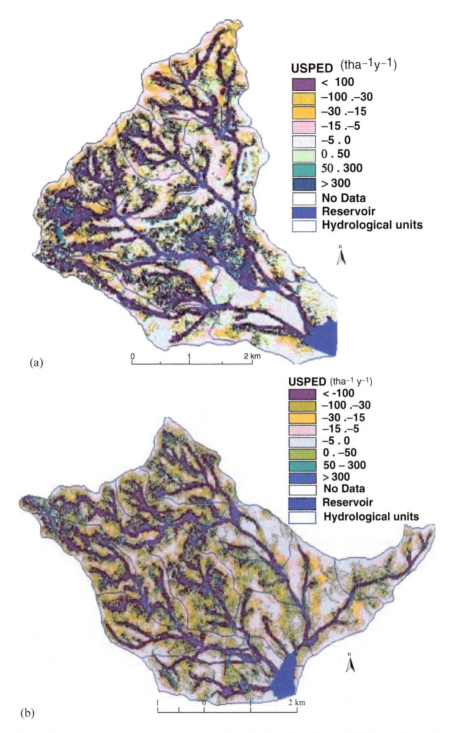

Fig. 5.10 Spatial patterns of erosion and deposition for (**a**) Adikenafiz and (**b**) Gerebmihiz catchments based on the USPED model in Tigray, northern Ethiopia. Negative values show erosion and positive values show deposition

The landscape positions where erosion is above the tolerable limit are located on the upslope, with steepness generally greater than 15°. These areas in most cases are characterized by convex slopeform with accelerated flow, which can facilitate soil loss and ultimate delivery to downslope positions. Higher elevations and steep-slope areas with poor surface cover and erodible lithology are more vulnerable to accelerated erosion than the lower slope areas of similar cover and lithologic attributes. Landscape positions where soil loss is within the tolerable limit are usually on low slope gradients of less than 8° with slightly higher erosion rates on culti-vated fields. However, widespread and collapsing gullies, which experience high soil loss, are located in the lower positions of catchments where the slopes are not very steep (Fig. 5.10). It is also observed that all steep-slope areas do not necessarily contribute high quantities of sediment compared to the gentle slope areas. Intensive land use with high soil disturbance is generally located on gentle slopes, while less intensive land use occurs on steeper slopes; in which case, the separate effects of both topography and land use on catchment response become less evident (Rustomji and Prosser, 2001). Some of the upslope positions with steep slopes have less erod-ible cliffs with little soil material to be detached and transported. Most steep-slope areas also have resistant lithology and good surface cover due to inaccessibility for cultivation and livestock grazing resulting in lower soil losses. The lower positions of catchments and piedmont sides are susceptible to rill/gully erosion as they are intensively cultivated and overgrazed.

The potential high soil erosion risk areas identified by the USPED model were evaluated during an independent field survey and with soil profile data. The results showed that the model generally correctly identified "hot-spot" areas of high soil erosion potential. This has made it possible to identify the landscape positions that are more vulnerable to erosion and therefore require prior conservation planning. Such information can help provide simplified information for decision-makers and planners on where intervention is necessary to reduce soil loss from catchments.

5.5   A Land-Use Planning Approach to Tackle Soil Erosion in Northern Ethiopia

The factor that can be manipulated to reduce the rate of soil loss from catchments and its downstream delivery is land-use/land-cover (LUC). Appropriate land-use and land-management practices that maintain good ground cover are useful means to reduce soil loss and sediment delivery (Erskine and Saynor, 1995). Since runoff and soil loss are both inversely related to ground cover (Costin, 1980; Nearing et al., 2005), afforestation could increase surface-roughness values and reduce the impact of raindrops and the ability of running water to detach and transport sedi-ments (Laflen et al., 1985; Vought et al., 1995). Afforestation of upslope and enhancement and maintenance of buffer riparian strips could offer tremendous benefits with respect to reducing on-site soil loss and off-site siltation of water bodies (Erskine and Saynor, 1995).

Model-based spatial scenarios can be used to simulate ways of preventing soil erosion and its downstream delivery (e.g., designing alternative land-use and conservation options targeted at specific locations) (Hessel et al., 2003). We used the USPED model in a GIS environment to simulate the effect of different land use change design measures on annual soil loss from catchments. The simulations mainly focused on reorganizing LUC types across different landscapes based on predefined criteria such as gullies, slope, and intensity of erosion. Five different simulations were performed and the net soil-loss rates were compared with those estimated for the baseline condition. The scenarios were run for two catchments (Adikenafiz and Gerebmihiz) in the Tigray region, northern Ethiopia.

5.5.1 Scenario Description

Before simulation, the catchments were divided into grid cells of equal size (10 m), and relevant catchment attributes were assigned. Erosion factors such as soil erodibility (K), cover and management (C), and support practice (P) were derived for each grid cell based on Hurni (1985) and Machado et al. (1996). The rainfall erosivity (R) factor was calculated considering the average rainfall of 35 years based on Hurni (1985). Terrain-related factors were calculated using the equation $(A_s)^m (\sin \beta)^n$, with A_s unit-contributing area, $\sin \beta$ slope angle, m and n coefficients. The USPED model was then used in a GIS to designate where we should intervene to reduce soil loss to a reasonable and acceptable level. The C and P factors were varied for each simulation after identifying areas/landscape positions of prior management intervention. The details of the different scenarios (Table 5.3) are as follows:

During the initial scenario (Table 5.3), the status quo net annual soil loss was determined using erosion factors that represent the current conditions of soils, LUC, rainfall, and management activities. The result was then used as a benchmark against which the result of each simulation was compared.

Scenario 2 is aimed at reducing soil loss from gullies and trapping sediment that moves along gully channels. Checkdams across gullies and grass buffers alongside are often considered cost-effective measures for reducing the sediment delivery to streams (Borin et al., 2005). Buffer vegetation, especially grass, acts as a filter by increasing surface roughness ultimately reducing the transport capacity of runoff and encouraging

Table 5.3 Summary of LUC-design and conservation scenarios for the Adikenafiz and Gerebmihiz catchments in Tigray, northern Ethiopia

Scenario	Description
1	Current/existing condition
2	25-m gully buffer terraced and grassed
3a	Areas over 25% slope enclosed/protected
3b	2 and 3a measures combined
4a	Areas experiencing soil loss of more than 25 t ha^{-1} year^{-1} enclosed
4b	2 and 4a measures combined

sediment deposition in the bufferstrip. To achieve this, gullies and their 25 m wide buffers were terraced (P factor 0.6)[1] and seeded with dense grass (C factor 0.01),[2] forming a stable grassed waterway to the reservoirs (Verstraeten et al., 2002).

Scenario 3 was applied to areas with slopes of more than 25%. These slopes were converted to enclosures (C factor = 0.01) and protected from human and livestock intervention to regenerate. The threshold slope class was defined according to the premise that slopes steeper than 25% will be less suitable for agriculture, so excluding them from cultivation may not have a significant impact on crop yield and livelihoods of farmers. In addition to enclosing areas with slope steeper than 25%, a simulation was run with terraced and grassed gullies (Scenario 3b) to assess their combined effect on net soil loss reduction.

Scenario 4a targeted areas experiencing high soil loss. Soil loss of more than 25 t ha^{-1} year^{-1} was selected as a threshold value for categorizing "hot-spot" (high soil erosion risk) areas. Note that for the region the maximum tolerable soil loss is around 18 t ha^{-1} year^{-1} (Hurni, 1985) and soil formation rates are around 6 t ha^{-1} year^{-1} (Hurni, 1983a, b); therefore soil loss of more than 25 t ha^{-1} year^{-1} is considered unacceptable. In Scenario 4a, the areas experiencing a soil loss rate of higher than the 25 t ha^{-1} year^{-1} threshold (Fig. 5.11) were enclosed to allow them to regenerate to dense grass/bush cover (C factor = 0.01) and the net soil loss after such intervention was calculated. An additional Scenario (4b) that considered conservation of gullies and vegetate erosion-prone areas of more than 25 t ha^{-1} year^{-1} with dense grass/bush was also applied to assess soil-loss reduction due to these integrated interventions.

5.5.2   *Soil Loss Reduction in Relation to the Different Interventions*

The soil-loss reduction as a result of each scenario is presented in Table 5.4. The results of the two catchments are discussed together. The average annual rate of soil loss from the two studied catchments based on scenario one was over 45 t ha^{-1} year^{-1} (Table 5.4). This is the net soil loss that can be observed under the existing catchment conditions and management practices. This rate is well above the tolerable soil loss and rate of soil generation in the region.

Scenario 2 shows that the rate of annual soil loss could be reduced by over 50% through conservation of gullies, exclusion of livestock, and avoidance of cultivation up to the very edge of gully banks. This reduction could be accomplished with less than 5% of the agricultural land left as set-aside. Scenario 3a indicates that when

---

[1] P factor values for support practices were defined for Ethiopia by Hurni (1985). Important values include ploughing up and down = 1.0; ploughing on contour = 0.9; strip cultivation = 0.8; terraces = 0.6; protected areas = 0.5.

[2] C factor values for Ethiopia were defined by Hurni (1985) for different cover types. Important values include Ethiopian **Teff** (*Eragrostis tef*) = 0.25; cereals/pulses = 0.15, sorghum/maize = 0.10; bush/shrub = 0.02; dense grass = 0.01; dense forest = 0.001.

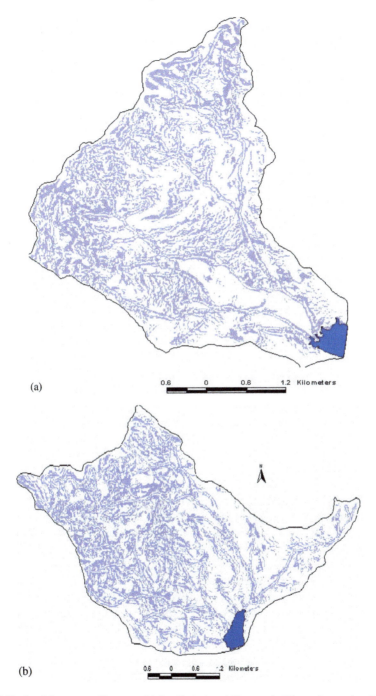

Fig. 5.11 Spatial patterns of hot-spot (high soil erosion risk) areas (soil loss of more than 25 t ha^{-1} year^{-1}) for two example catchments (**a**) Adikenafiz and (**b**) Gerebmihiz in Tigray, northern Ethiopia

Table 5.4 Net soil loss, proportion of cultivable land to be set aside, and areas to be enclosed during each scenario for two catchments in Tigray, northern Ethiopia

	Adikenafiz			Gerebmihiz		
Scenario	Net soil loss (t ha^{-1} year^{-1})	Cultivated land foregone[*]	Area to be enclosed[**]	Net soil loss (t ha^{-1} year^{-1})	Cultivated land foregone[*]	Area to be enclosed[**]
1	51	n.a.	n.a.	43	n.a.	n.a.
2	28	4	—	18	3	—
3a	47	10	25	35	8	14
3b	26	15	19	15	10	13
4a	27	27	38	19	24	35
4b	20	30	30	11	29	35

[*]Indicates the proportion of cultivated land (%) to be set aside and enclosed.
[**]Indicates the proportion of land (%) to be afforested (enclosed), excluding grassed gully buffers.

areas with slopes greater than 25% are enclosed, a soil-loss reduction of about 13% can be achieved compared to the current situation. When conservation of gullies is combined with enclosure of areas with steeper slopes (Scenario 3b), the net soil loss reduction improves to over 55% (Table 5.4).

When areas experiencing soil loss greater than 25 t ha^{-1} year^{-1} were enclosed and allowed to regenerate to dense cover (Scenario 4a), a soil loss reduction of about 50% could be achieved. Through integrated management of erosion-sensitive areas and conservation of gullies (Scenario 4b), the soil loss reduction was about 65% (Table 5.4). This intervention improves the soil-loss reduction by about15% compared to Scenario 4a.

Generally, conservation practices that involve gullies show relatively better impact in reducing soil loss from catchments. This is because gullies are very prominent features in the two study sites, serving as prime sediment sources and providing catchment connectivity. Management of gullies reduces soil loss better in the Gerebmihiz catchment than in the Adikenafiz catchment (Table 5.4); this is mainly because the USPED model predicted a higher level of gully erosion in the former than in the latter. Enclosing steep-slope areas so as to increase frictional resistance by dense cover shows a relatively low reduction in soil loss (Scenario 3a) compared to that which can be achieved when targeting gullies. This may be because upslope areas are predominantly covered with resistant and less erodible rocks with low sediment-yield potential and therefore are not prime sediment sources. In addition, most steep-slope areas are less accessible and not very much exposed to disturbances by livestock and humans. When integrated management of steep-slope areas and gullies (Scenario 3b) is applied, the soil-loss reduction improved significantly (Table 5.4). Conservation measures targeting areas of high soil-loss risk (Scenario 4a) almost doubled the potential soil-loss reduction compared with that which can be achieved by enclosing steep-slope areas alone (Scenario 3a). This shows that steep-slope areas do not necessarily experience high soil loss. The soil-loss reduction improves further when both hot-spot areas of soil loss and gullies are protected (Scenario 4b).

5.5.3 *Preliminary Evaluation of the Efficiency of the Different Scenarios*

The different scenarios show a range of soil-loss reduction achievements depending upon the locations and types of interventions. The choice of the best measure(s) can be made by roughly comparing the benefits due to soil-loss reduction to the costs mainly due to exclusion of arable lands from cultivation. Since such cost–benefit analyses were not conducted in our study, the amount of soil-loss reduction was compared with the respective proportion of cultivated land to be set aside and enclosed.

Management interventions related to the conservation of gullies (Scenario 2) can be considered relatively effective because the fraction of cultivable land to be set aside is low while the potential reduction in soil loss is high (Table 5.4). Scenario 3b (conservation of sloping areas and gullies) could also be considered efficient since the fraction of cultivated land to be withdrawn from its current use and enclosed is relatively small compared with the associated high potential soil loss reduction that can be achieved. According to Scenario 3b, a potential sediment-yield reduction of about 55% could be achieved by withdrawing about 15% of cultivated land from its current use and enclosing it. This is a far better than Scenario 3a (conservation of steep slopes), which requires similar fractions of cultivated land to be set aside but delivers only a small change in net soil-loss reduction (Table 5.4).

Scenarios 4a and 4b (management of high soil-loss risk areas and gullies) showed significant reduction in sediment yield. However, large parts (about 30%) of the cultivated land of the catchments will need to be enclosed and protected from livestock and human interference to achieve about 60% soil-loss reduction (Table 5.4). While the amount of soil-loss reduction is very high, exclusion of over 30% of the cultivated land from its current use may have a significant impact on local farmers, at least in the short term. The efficiency of the two scenarios may therefore be limited from a practical point of view.

Generally, simple comparison of the proportion of cultivable land to be set aside with the proportion of soil-loss reduction shows that Scenarios 2 and 3b are the most efficient. However, detailed assessment of the trade-offs of each landscape position in terms of its productivity and conservation requirement need to be conducted to evaluate whether setting aside a given fraction of land is worthwhile. For instance, if the areas to be set aside have limited agricultural productivity, enclosing them may be an appropriate and efficient intervention despite the proportion of the areas to be set aside. In most cases, areas where high soil loss is experienced are either no longer under cultivation or are not properly managed and do not have a high productivity. Excluding such areas from cultivation may not have a significant effect on the livelihoods of farmers. Since enclosing areas of high soil-loss risk could exclude large proportion of cultivated land, it may be necessary to analyze the soil-loss reduction potential of other alternative conservation practices such as vegetative barriers across slopes.

Most scenarios show that a relatively large decrease in the proportion of cultivated land may be required (except for the case of gullies) to achieve a reasonable decrease in soil loss from catchments. This may not be acceptable considering the already small plots of land owned by farmers in the study area. The decrease in arable land should, therefore, be accompanied by an intensification of the remaining cropland and by an increase in benefits from the proposed enclosures. Since it takes time before the new land use can start to benefit the farmers, the government may need to consider mechanisms of compensation to make the change economically feasible for farmers in the short term, until they are able to derive goods and services from the enclosures. Detailed socioeconomic impact assessment would therefore be necessary to understand the potentials and drawbacks of the management scenarios. In addition to the above-mentioned, land-use planning-based scenarios, it is also important to incorporate indigenous soil and water conservation practices because imposing only external "solutions" without regard for local practices may not be successful and sustainable (Reij et al., 1996).

5.6 Conclusion

Soil erosion is a global issue that threatens human livelihoods and civilization. A number of factors contribute to the high rates of soil loss at global, regional, and local levels. Climate, slope, soils, surface cover, and land management are the most important. Land misuse, deforestation, and overgrazing lead to severe erosion and excessive sediment yield. Population pressure leads to use of marginal lands and steep slope which could accelerate erosion processes.

In the drylands of Ethiopia, a combination of natural and human factors aggravates soil erosion. Additionally, the physical makeup of the highlands with gorges and other topographic barriers restricts the implementation of management and conservation practices. With increasing numbers of people, cultivation has expanded into ecologically fragile and marginal mountainous lands. Soil erosion therefore increases, and is a symptom of misuse and mismanagement. To deal with erosion, site-specific management measures are necessary.

The statistical analyses employed in this study enabled us to understanding the major determinant factors that provoke and accelerate soil erosion in the highlands of northern Ethiopia. The modeling exercise makes it possible to identify hot-spot areas of erosion that require urgent intervention. The land-use planning-based spatial scenarios demonstrate the possibility of reducing soil loss from catchments through the implementation of a range of management activities. In summary, the quantitative sediment-yield data acquired for the selected catchments enabled us to answer an important question: "How severe is the problem and which sites in the region require priority attention?" The statistical analysis based on soil-loss data and corresponding catchment attributes helped answer the question: "Which environmental attributes of catchments accelerate soil erosion and therefore need priority attention?" The distributed modeling approach enabled us to identify and flag landscape positions within catchments which experience high soil-loss rates. Finally, the spatial simulations helped us to assess "what conservation measures were more efficient and where they are best sited." It would be particularly useful to go through the modeling scheme we have discussed prior to the construction of dams for water harvesting or energy generation.

Acknowledgments The expenses related to field data collection were covered by DAAD (Germany), Center for Development Research (ZEF, Germany), and Mekelle University (Ethiopia).

References

Aksoy, H., & Kavvas, M. L., (2005). A review of hillslope and watershed scale erosion and sediment transport models. *Catena, 64*, 247–271.

Borin, M., Vianello, M., Morari, F., & Zanin, G. (2005). Effectiveness of buffer strips in removing pollutants in runoff from cultivated field in North-East Italy. *Agriculture, Ecosystems and Environment, 105*, 101–114.

Costin, A. B. (1980). Runoff and soil nutrient losses from an improved pasture at Ginninderra, Southern Tablelands, New South Wales. *Australian Journal of Agricultural Research, 31*, 533–546.

Dearing, J. A., & Foster, D. L. (1993). Lake sediments and geomorphological processes: Some thoughts. In J. McManus & R. W. Duck (Eds.), *Geomorphology and sedimentology of lakes and reservoirs* (pp. 73–92). Chichester, UK: Wiley.

De Roo, A. P. J. (1998). Modelling runoff and sediment transport in catchments using GIS. *Hydrological Processes, 12*, 905–922.

den Biggelaar, C., Lal, R., Wiebe, K., Eswaran, H., Breneman, V., & Reich, P. (2004). The global impact of soil erosion on productivity II: Effect on crop yields and production over time. *Advances in Agronomy, 81*, 49–95.

Desmet, P. J. J., & Govers, G. (1996). Comparison of routing algorithms for digital elevation models and their implications for predicting ephemeral gullies. *International Journal of Geographical Information Systems, 10*, 311–331.

Dickinson, A., & Collins, R. (1998). Predicting erosion and sediment yield at the catchment scale. In F. W. T. Penning de Vries, F. Agus & J. Kerr (Eds.), *Soil erosion at multiple scales: Principles and methods for assessing causes and impacts* (pp. 317–342). Wallingford, UK: CABI, in association with the International Board for Soil Research and Management.

Dregne, H. E. (1990). Erosion and soil productivity in Africa. *Journal of Soil and Water Conservation, 45*(4), 431–436.

El-Swaify, S. A. (1994). State-of-the-art for assessing soil and water conservation needs and technologies. In T. L. Napier, S. M. Camboni & S. A. El-Swaify (Eds.), *Adopting conservation on the farm* (pp. 13–27). Ankeny, IA: Soil and Water Conservation Society of America.

Erskine, W. D., & Saynor, M. J. (1995). *The influence of waterway management on water quality with particular reference to suspended solids, phosphorus and nitrogen.* Victoria, East Melbourne: Department of Conservation and Natural Resources.

Eswaran, H., Lal, R., & Reich, P. F. (2001). Land degradation: An overview. In E. M. Bridges, I. D. Hannam, L. R. Oldeman, F. W. T. Pening de Vries, S. J. Scherr & S. Sompatpanit (Eds.), *Responses to land degradation. Proceedings of 2nd International Conference on Land Degradation and Desertification. Khon Kaen, Thailand.* New Delhi, India: Oxford Press.

Eweg, H., & Van Lammeren, R. (1996). *The application of a GIS at the rehabilitation of degraded and degrading areas. A case study in the highlands of Tigray, Ethiopia.* Wageningen, The Netherlands: Centre for Geographical Information Processing, Agricultural University.

FAO. (1986). *Ethiopian highland reclamation study.* Ethiopia. Final report. Rome: FAO.

Fargas, D., Martínez-Casasnovas, J. A., & Poch, R. M. (1997). Identification of critical sediment source areas at regional level. *Physics and Chemistry of the Earth, 22*, 355–359.

Foster, I. D. L. (1995). Lake and reservoir bottom sediments as sources of soil erosion and sediment transport data in the UK. In I. D. L. Foster, M. M. Gurnell & B. Webb (Eds.), *Sediment and water quality in river catchments* (pp. 265–283). Chichester, UK: Wiley.

Grayson, R., & Blöschl, G. (2000). Spatial modelling of catchment dynamics. In R. Grayson & G. Blöschl (Eds.), *Spatial patterns in catchment hydrology: Observations and modelling*. Cambridge: Cambridge University Press.

Harden,C. P. (1993). Land use, soil erosion, and reservoir sedimentation in an Andean drainage basin in Ecuador. *Mountain Research and Development, 13*, 177–184.

Hessel, R., Messing, I., Liding, C., Ritsema, C., & Stolte, J. (2003). Soil erosion simulations of land use scenarios for a small Loess Plateau catchment. *Catena, 54*, 289–302.

Hurni, H. (1983a). *Soil formation rates in Ethiopia*. Working paper 2. Addis Ababa: Ethiopian Highlands Reclamation Studies.

Hurni, H. (1983b). Soil erosion and soil formation in agricultural ecosystems in Ethiopia and Northern Thailand. *Mountain Research and Development, 3*,131–142.

Hurni, H. (1985). Erosion-productivity-conservation systems in Ethiopia. Soil Conservation Research Project (SCRP). In I. P. Sentis (Ed.), *Soil conservation and productivity* (pp. 654–674). Proceedings of 4th International Conference on Soil Conservation, Venezuela.

Hurni, H. (1990). Degradation and conservation of soil resources in the Ethiopian highlands. *Mountain and Research Development, 8*, 123–130.

Hurni, H. (1993). Land degradation, famine, and land resource scenarios in Ethiopia. In D. Pimentel (Ed.), *World soil erosion and conservation*. Cambridge, Cambridge University Press.

Hurni, H., & Perich, I.(1992). *Towards a Tigray regional environmental and economic strategy (TREES): A contribution to the symposium on combating environmental degradation in Tigray, Ethiopia*. Berne: Group for Development and Environment, Institute of Geography, University of Bern, Switzerland.

Jetten, V., Govers, G., & Hessel, R. (2003). Erosion models: Quality of spatial predictions. *Hydrological Processes, 17*, 887–900.

Jiang, D., Qi, L., & Tan, J. (1980). Soil erosion and conservation in the Wuding River Valley, China. In R. P. C. Morgan (Ed.), *Soil conservation: Problems and prospects* (pp. 461–479). Chichester, UK: Wiley.

Julien, P. Y., & Simons, D. B. (1985). Sediment transport capacity of overland flow. *Transactions of the American Society of Agricultural Engineers, 28*, 755–762.

Kirkby, M. J., Le Bissonais, Y., Coulthard, T. J., Daroussin, J., & McMahon, M. D. (2000). The development of land quality indicators for soil degradation by water erosion. *Agriculture, Ecosystems and Environment, 81*, 125–136.

Laflen, J. M., & Roose, E. (1998). Methodologies for assessment of soil degradation due to water erosion. In R. Lal, W. H. Blum, C. Valentine & B. A. Stewart (Eds.), *Advances in soil science* (pp. 31–55). New York: CRC Press.

Laflen, J. M., Foster, G. R., & Onstad, C. A. (1985). Simulation of individual-storm soil loss for modeling the impact of soil erosion on crop productivity. In S. A. El-Swaify, W. C. Moldernhauer & A. Lo (Eds.), *Soil erosion and conservation* (pp. 285–295). Ankeny, IA: Soil and Water Conservation Society of America.

Lal, R. (1994). Soil erosion by wind and water: Problems and prospects. In R. Lal (Ed.), *Soil erosion research methods* (pp. 1–9). Ankeny, IA: Soil and Water Conservation Society of America.

Lal, R. (1995). Erosion-crop productivity relationship for soils in Africa. *American Journal of Social Science Society, 59*, 661–667.

Lane, L. J., Renard, K. G., Foster, G. R., & Laflen, J. M. (1997). Development and application of modern soil erosion prediction technology: The USDA experience. *Eurasian Soil Science, 30*(5), 531–540.

Lawrence, P., & Dickinson, A. (1995). *Soil erosion and sediment yield: A review of sediment data from rivers and reservoirs* (Report prepared under FAO writers' contract). Wallingford, UK: Overseas Development Unit, HR Wallingford Ltd.

Lu, X. X., & Higgitt, D. L. (1999). Sediment yield variability in the Upper Yangtze, China. *Earth Surface Processes and Landforms, 24*, 1077–1093.

Machado, M. J., Perez-Gonzalez, A., & Benito, G. (1996). Assessment of soil erosion using a predictive model. In E. Feoli (Ed.), *Rehabilitation of degrading and degraded areas of Tigray, Northern Ethiopia* (pp. 237–248). Trieste, Italy: Department of Biology, University of Trieste.

Maidment, D. R. (1996). Environmental modeling with GIS. In M. F. Goodchild, L. T. Steyaert, B. O. Parks, C. Johnston, D. Maidment & S. Glendinning (Eds.), *GIS and environmental modeling: CO, Progress and Research Issues* (pp. 315–323). Fort Collins, CO: GIS World.

Merritt, W. S., Letcher, R. A., & Jakemna, A. J. (2003). A review of erosion and sediment transport models. *Environmental Modelling and Software, 18*, 761–799.

Mitas, L., & Mitasova, H. (1998). Distributed soil erosion simulation for effective erosion prevention. *Water Resources Research, 34*, 505–516.

Mitasova, H., Hofierka, J., Zloch, M., & Iverson, L. R. (1996). Modelling topographic potential for erosion and deposition using GIS. *International Journal of Geographical Information Systems, 10*, 629–641.

Mitasova, H., Mitas, L., & Brown, W. M. (2001). Multiscale simulation of land use impact on soil erosion and deposition patterns. In D. E. Stott, R. H. Montar & G. C. Steinhardt (Eds.), *Sustaining the global farm*. The 10th International Soil Conservation Organization (24–29, 1999). West Lafayette, IA: Purdue University and the USDA-ARS National Soil Erosion Research Laboratory.

Moore, I. D., Grayson, R. B., & Ladson, A. R. (1991). Digital terrain modelling: A review of hydrological, geomorphological and biological applications. *Hydrological Processes, 5*, 3–30.

Moore, I.D., Wilson, J.P., & Ciesiolka, C.A. (1992). Soil erosion prediction and GIS: Linking theory and practice. In S. H. Luk & J. Whitney (Eds.), *Proceedings of the International Conference on the application of geographical information systems to soil erosion management* (pp. 31–48). Toronto: University of Toronto Press.

Morgan, R. P. C. (1995). *Soil erosion and conservation*, 2nd ed. New York: Wiley.

Nearing, M. A., Jetten, V., Baffaut, C., Cerdan, O., Couturier, A., Hernandez, M., Le Bissonnais, Y., Nichols, M. H., Nunes, J. P., Renschler, C. S., Souchere, V., & van Oost, K. (2005). Modeling response of soil erosion and runoff to changes in precipitation and cover. *Catena 61*, 131–154.

Nyssen, J. (2001). *Erosion processes and soil conservation in a tropical mountain catchment under threat of anthropogenic desertification – a case study from Northern Ethiopia*. PhD Thesis. Faculteit Wetenschappen. Department Goegrafie – Geologie, Katholieke University Leuven, Belgium.

Oldeman, L.R. (1994). The global extent of soil degradation. In D. J. Greenland & I. Szabolcs (Eds.), *Soil resilience and sustainable land use*. Wallingford, UK: CABI.

Phippen, S., & Wohl, E. (2003). An assessment of land use and other factors affecting sediment loads in the Rio Puerco watershed, New Mexico. *Geomorphology, 52*, 269–287.

Pimentel, D., Harvey, C., Resosudarmo, P., Sinclair, K., Kurz, D., McNair, M., Crist, S., Shpritz, L., Fitton, L., Saffouri, R., & Blair, R. (1995). Environmental and economic costs of soil erosion and conservation benefits. *Science, 267*, 1117–1123.

Poesen, J., Nachtergaele, J., Verstraeten, G., & Valentin, C. (2003). Gully erosion and environmental change: Importance and research needs. *Catena, 50*, 91–133.

Reij, C., Scoones, I., & Toulmin, C. (1996). *Sustaining the soil: Indigenous soil and water conservation in Africa*. London: Earthscan.

Renard, K. G., & Foster, G. R. (1983). Soil conservation: Principles of erosion by water. In H. E. Degne & W. O. Willis (Eds.), *Dryland agriculture. Agronomy Monograph, 23*, 156–176. Soil Science Society of America, Madison, WI.

Rustomji, P., & Prosser, I. (2001). Spatial patterns of sediment delivery to valley floors: Sensitivity to sediment transport capacity and hillslope hydrology relations. *Hydrological Processes, 15*, 1003–1018.

Sertsu, S. (2000). *Degraded soils of Ethiopia and their management*. Proceedings of the FAO/ISCW expert consultation on management of degraded soils in Southern and East Africa. 2nd Network Meeting, 18–22 September 2000. Pretoria.

Shibru, D., Rieger, W., & Strauss, P. (2003). Assessment of gully erosion using phtotogrammetric techniques in Eastern Ethiopia. *Catena, 50*, 273–291.

Stott, A. P., Butcher, D. P., & Pemberton, T. J. L. (1988). Problems in the use of reservoir sedimentation data to estimate erosion rates. *Zeitschrift für Geomorphologie, 30,* 205–226.

Taddese, G. (2001). Land degradation: A challenge to Ethiopia. *Environmental Management, 27,* 815–824.

Tamene, L., Park, S., Dikau, R., & Vlek, P. L. G. (2006). Analysis of factors determining sediment yield variability in the highlands of Northern Ethiopia. *Geomorphology, 76,* 76–91.

Tilahun, Y., Esser, K., Vägen, T. G., & Haile, M. (2002). *Soil conservation in Tigray, Ethiopia.* Norway: Äs, Agricultural University of Norway, Noragric, Report No.5.

Trimble, S. W. (1974). *Man-induced soil erosion in the southern Piedmont, 1700–1970.* Ankeny, IA: Soil Conservation Society of America.

Verstraeten, G., & Poesen, J. (2001). Factors controlling sediment yield from small intensity cultivated catchments in a temperate humid climate. *Geomorphology, 40,* 123–144.

Verstraeten, G., Van Oost, K., Van Rompaey, A., Poesen, J., & Govers, G. (2002). Evaluating an integrated approach to catchment management to reduce soil loss and sediment pollution through modelling. *Soil Use and Management, 19,* 386–394.

Vought, L. B., Pinay, G., Fuglsang, A., & Ruffinoni, C. (1995). Structure and function of buffer strips from a water quality perspective in agricultural landscapes. *Landscape and Urban Planning, 31,* 323–331.

Walling, D. E., & Webb, D. W. (1981). The reliability of suspended sediment load data: Erosion and sediment transport measurement. Proceedings of the Florence Symposium, Florence. *International Association of Hydrological Sciences,* 177–194.

Wilson, J., & Gallant, C. (2000). Digital terrain analysis. In J. Wilson & J. Gallant (Eds.), *Terrain analysis: Principles and applications.* New York: Wiley.

Wilson, J., & Lorang, M. S. (1999). Spatial models of soil erosion and GIS. In A. S. Fotheringham & M. Wegener (Eds.), *Spatial models and GIS: New potential and new models.* London: Taylor & Francis.

Chapter 6
Soil Degradation Under Irrigation

Paul L.G. Vlek, Daniel Hillel, and Ademola K. Braimoh

Abstract Irrigation is a precondition for stable crop production in areas characterized by marked variability in rainfall distribution. Despite substantial investment in irrigation projects in the past decades, global irrigated cropland area has hardly grown. Here we discuss the factors related to the sustainability of irrigation and strategies to alleviate them. Water resources deterioration, diversion of water for other uses, and soil degradation are the major factors affecting the environmental sustainability of irrigated agriculture. Water logging results from the tendency to apply water in excess of irrigation requirements. It leads to reduced aeration, nutrient uptake, and crop yields. Salt buildup occurs through the process of capillary rise when the water table rises close to the surface. Salinity risks also increases when saline water is used for irrigation and when poor fertilizer and poor irrigation management are combined. Groundwater drainage is the ultimate precautionary measure against groundwater rise and salt accumulation, but its timely installation is essential for optimal result. Constant soil monitoring and the use of systematic diagnosis testing could also reveal the incipience of soil salinity. Future global warming would likely exacerbate water demand for irrigation with the implications that crops would grow in hotter, drier, and more saline conditions. The ability of irrigated agriculture to meet future challenges would therefore depend on the progress of new research to enhance adaptation to these changes.

Keywords Soil salinity, waterlogging, drainage, salt tolerance, climate change

6.1 Introduction

Irrigation, the supply of water to crops by artificial means is *sine qua non* for stable agricultural production both in arid regions and areas characterized by marked seasonal variability in rainfall. By supplementing the limited natural supply of water, irrigation helps to raise crop yields. It also aids land-use intensification and crop diversification in dryland areas by prolonging the effective growing period, thereby permitting multiple cropping (2–4 crops) per year. Lastly, irrigation provides synergy

A.K. Braimoh and P.L.G. Vlek (eds.), *Land Use and Soil Resources.*   101
© Springer Science + Business Media B.V. 2008

with other intensification inputs such as pesticides, fertilizers, and improved crop varieties whose efficiency depends on water availability. Therefore, irrigation provides incentives for farmers to invest in agricultural intensification inputs.

Water is a major constraint to food production in several parts of the world. With the exception of few areas, the rapid growth in food production during the green revolution occurred mainly on irrigated lands. Aggregate food supply in Asia more than doubled between 1970 and 1995, with only a 4% increase in the net cropland area (Rosegrant and Hazell, 1999). Average cereal yields under irrigation are typically twice those obtained under rainfed conditions, and irrigation continues to play a major role in feeding the world's growing population, contributing over 40% to total agricultural production.

Of all the major human activities, agriculture remains the largest user of water at the global level, accounting for about 70% of total freshwater withdrawal (Table 6.1). With the exception of Europe and North America that are relatively more industrialized, agricultural sector is the largest source of freshwater withdrawal in all regions of the world. In 2001, agricultural uses accounted for about 5%, 10%, and 17% of the internal renewable water resources of Africa, the Caribbean, and Asia, respectively. Asia has the largest proportion of global freshwater withdrawal for agriculture of about 73%.

In developing countries, a substantial proportion of investment in agriculture is usually from domestic sources, and irrigation is often the largest beneficiary of public agricultural investment. Table 6.2 shows that irrigated croplands increased in all the world regions between 1980 and 2003, suggesting significant investment in irrigation in the last 2 decades. Between 1980 and 1990, the amount of land equipped with

Table 6.1 Freshwater withdrawal by different sectors across world regions in 2001. (Based on FAO Aquastat database available at http://www.fao.org/ag/agl/aglw/aquastat/main/index.stm)

Continent/ Region	IRWR* km³/year	Total volume of freshwater utilization km³/year	Freshwater withdrawal by sector						Total withdrawal as % of IRWR
			Domestic		Industrial		Agricultural		
			km³/year	%	km³/year	%	Km³/year	%	
Africa	3 936	215	21	10	9	4	184	86	**5.5**
Asia	11 594	2 378	172	7	270	11	1 936	81	**20.5**
Latin America	13 477	252	47	19	26	10	178	71	**1.9**
Caribbean	93	13	3	23	1	9	9	68	**14.4**
North America	6 253	525	70	13	252	48	203	39	**8.4**
Oceania	1 703	26	5	18	3	10	19	72	**1.5**
Europe	6 603	418	63	15	223	53	132	32	**6.3**
World	**43 659**	**3 830**	**381**	**10**	**785**	**20**	**2 664**	**70**	**8.8**

*Internal Renewable Water Resources.

Table 6.2 Regional distribution of area equipped for irrigation based on FAO Aquastat. (Database available at http://www.fao.org/ag/agl/aglw/aquastat/main/index.stm)

Continent/Region	Area (million ha)			As % of cropland		
Year	1980	1990	2003	1980	1990	2003
Africa	9.5	11.2	13.4	5.1	5.7	5.9
Asia	132.4	155.0	193.9	28.9	30.5	34.0
Latin America	12.7	15.5	17.3	9.4	10.9	11.1
Caribbean	1.1	1.3	1.3	16.4	17.9	18.2
North America	21.2	21.6	23.2	8.6	8.8	9.9
Oceania	1.7	2.1	2.8	3.4	4.0	5.4
Europe	14.5	17.4	25.2	10.3	12.6	8.4
World	193.0	224.2	277.1	15.8	17.3	17.9

irrigation increased substantially in Asia (by 17%), Africa and the Caribbean (by 18%), Europe (by 20%), Latin America (by 22%), and Oceania (by 26%). With the exception of Latin America and the Caribbean, the increase between 1990 and 2003 was higher than the increase in the preceding decade (1980–1990). Europe, Oceania, and Asia recorded the highest increase of 45%, 34% and 25%, respectively between 1990 and 2003. At the global level, the area equipped for irrigation increased by 18% between 1980 and 1990, and by 21% between 1990 and 2003.

Some estimates indicate that annual investment in irrigation in developing countries range from US$60–80 billion (DFID, 2000; World Bank 2001). The marginal increase in total irrigated cropland (Table 6.2) is, however, not commensurate with this huge expenditure. Between 1980 and 1990, the increase in proportion of cropland area under irrigation was 6% for Asia, 12% for Africa, 9% for the Caribbean, 22% for Europe, 16% for Latin America, and 18% for Oceania. With the exception of Asia, North America, and the Oceania, the increase in proportion of cropland area under irrigation between 1990 and 2003 for other regions of the world was lower than the increase between 1980 and 1990. At the global level, the cropland area under irrigation increased by 12% between 1980 and 1990, and by just 5% between 1990 and 2003. Irrigated area amounts to only 18% of world's cropland area (Table 6.2). About 70% of the world's irrigated land is in Asia, whereas Africa has less than 5%. Thus, taken globally, despite the perceived increase in investments, the increase in total area under irrigation has hardly grown.

Two major reasons are responsible for the marginal increase in irrigated cropland. Reason one is the dilemma of water resource deterioration, that is, the depletion and/or pollution of water resources upon which irrigation projects are based. In many situations, water has been diverted into other uses leading to fragmentation of rivers, alteration of flow regimes, and change in flow and regime with profound ecological consequences. Furthermore, competition among many uses or among many administrative units affects the availability of water for irrigation. Lastly, escalating energy costs markedly affect the delivery of water for irrigation, making irrigated agriculture unprofitable. The second reason is the physical and chemical degradation of large expanse of irrigated land to the point where further cultivation is uneconomic or rehabilitation of degraded land is economically prohibitive.

This chapter addresses the problems affecting the sustainability of irrigation and measures necessary to alleviate them. Section 6.2 provides a summary of the upstream and downstream effects of dams which regulate flow of water for irrigated croplands. Section 6.3 examines the twin problems of waterlogging and soil salinity. The problems are so widespread and pervasive that many are pessimistic that no irrigation venture can be sustained in the long run. Leaching and soil amendment to support irrigation are discussed in Section 6.4, whereas Section 6.5 discusses groundwater drainage, an integral aspect of sustainable irrigation management. In Section 6.6, some studies are further reviewed to highlight irrigation problems in different locales. Lessons from the case studies generally reveal the need for early warning system to indicate the onset of water table rise and salt buildup and thereby respond accordingly. The prospects of global warming due to enhanced atmospheric greenhouse effects are discussed in Section 6.7. The possible intensification of the hydrologic cycle calls for adaptation strategies to cope with the expected changes. Section 6.8 concludes with imperatives for viability of irrigation projects.

6.2   Effects of Dam Construction

In irrigation projects, dams and canals are usually constructed to regulate flow of water to ensure adequate storage and supply throughout the growing season. Thus, dams represent a major interaction between humans and the hydrologic cycle in irrigation development. Currently, about half of the world's rivers have at least one large dam (higher than 15 m). In all, about 45,000–50,000 large dams have been built for various purposes ranging from hydropower to domestic, industrial, and agricultural uses. About 50% of the world's large dams were built primarily for irrigation, and some 30–40% of the world's irrigated cropland worldwide relies on dams (World Commission on Dams, 2000).

Dam construction often poses several environmental problems, as the ideal topographic, geologic, and climatic conditions are seldom found in the proximity of target irrigation projects. This siting of dams at unfavorable locations makes the construction and maintenance of dams economically prohibitive and environmentally unsustainable. The environmental impact of dams is summarized in Table 6.3. The effects generally result from inundation, flow manipulation, and fragmentation of river systems (Nilsson et al., 2005).

Apart from dam construction, another water engineering activity that may impact the environment is aquifer mining. Aquifer mining occurs in arid regions where insufficient surface water necessitates the use of groundwater for irrigation. Because the natural rate of water recharge is low, most of the water extracted from the groundwater pool is translocated from the aquifer to the atmosphere (Vorosmarty and Sahagian, 2000).

A global overview of dam-based impacts on 292 large river systems (with annual mean flow of at least $350 \, m^3 \, s^{-1}$) draining 54% of the world's land area (Nilsson et al., 2005) shows that nearly 139 of the large river systems remain unfragmented by dams in the main channel, 119 have unfragmented tributaries, and

Table 6.3 Upstream and downstream impacts of dams

Upstream	Field submergence
	Damage to natural ecosystems
	Displacement of human population could lead to changes in land-use pattern
	Resettlement of people could also create health problems
	Displacement of infrastructure
	Loss of cultural heritage and scenic sites
	Increased vulnerability to natural disasters in some locations
	Water loss through leaching and direct evaporation
	Siltation, leading to reduction in dam's capacity
	Accelerated erosion
Downstream	Reduction in river flow leads to deprivation of water for agriculture and other uses
	Water deprivation leads to desiccation and reduces biological diversity of riparian ecosystems
	Where a river naturally discharges into a freshwater lake, diversion of water for irrigation causes shrinkage of the lake (e.g., Aral Sea and Dead Sea)
	Irrigation water contains salts and fertilizer residues that pollutes the downstream

102 are completely unfragmented. The strongly affected river systems constitute 52% (i.e., about 41 million km^2) of total large river system catchment area, and include the 25 river systems with the highest irrigation pressure, 15 of which are situated in Asia. Most of the unaffected large river systems are situated in just four biomes (tundra, boreal forests, tropical, and subtropical moist broadleaf forests; tropical and subtropical grasslands; savannahs; and shrublands), whereas the catchment area of strongly affected river systems constitute more than 50% of three biomes (temperate broadleaf and mixed forests; temperate grasslands, savannahs, and shrublands; and flooded grasslands and savannahs).

Environmental damage and the question of financial viability caused by dams have led to increasing resistance to the construction of large dams by international donors such as the World Bank. Between 1950 and 1993, World Bank investments for irrigation projects was 7% of its total lending, but this decreased to about 4% between 1990 and 1997 (DFID, 2000). However, the need to increase food production will make irrigation expansion indispensable in the future. Small dams may not be able to cope with the water problems posed by increasing agricultural, domestic, and industrial demands. Thus, efforts to minimize the impacts of dams will continue to be a major challenge.

6.3 The Problem of Water Logging and Soil Salinity

From its inception in the Ancient Mesopotamia about 6,000 years ago, irrigation has induced two major processes of degradation that threatens its sustainability. The self-destructive, twin menaces of waterlogging and salinization that brought about the downfall of ancient hydraulic civilizations continue to plague irrigated agriculture today (Hillel, 2000).

6.3.1 Waterlogging

Waterlogging is a situation in which soil pores are filled with water due to low porosity or poor drainage. It usually results from the application of water in excess of crop requirements. It is common in areas where the following situation prevails: low-lying lands, fine-textured soils with low permeability, arid regions with high rates of evaporation, brackish water (containing 0.5–30 g of dissolved salts per liter) and high concentration of Na in irrigation water, inadequate provision for artificial drainage, and excessive tillage and soil compaction. In arid and semiarid regions, there is always the tendency to apply too much water, as natural drainage is hardly sufficient to leach out the salts introduced by irrigation water. Water flowing down the root zone eventually percolates into the water table, increases groundwater, and saturates the subsoil. The further addition of irrigation water especially when the water table is shallow causes the water table to rise. It is the saturation of the root zone with water that leads to reduced aeration, nutrient uptake, crop growth, and yield. Total crop failure is common for most crops as carbon dioxide accumulates to toxic level. The only remedy in most instances is to plant rice because of its ability to transmit oxygen from its leaves to the roots. Estimated waterlogging thresholds (as a function of water table depth) at which yields are negatively affected for selected crops is presented in Table 6.4.

As a result of poor irrigation practice, waterlogging occurs on about 60 million ha of irrigated land worldwide. Turkmenistan is an example of areas of the world where waterlogging threatens the sustainability of agriculture. Agriculture in Turkmenistan is entirely dependent on irrigation owing to its aridity: low annual rainfall (varying from 90 mm in Dashouz to about 400 mm in Kopet Dag) and high average temperatures (above 45 °C in the hottest months June–August). Due to government policies that placed priority on cotton, irrigated cropland in Turkmenistan increased from about 64,000 ha in 1884 to about 1.5 million ha in 1994. Expansion of irrigation network was associated with increased groundwater levels in all parts of Turkmenistan from 1986 to 1992 (O'Hara, 1997). The amount of land where groundwater levels were less than 2 m below the surface increased by over 285,000 ha of irrigated areas. The combined total of water losses in irrigation conveyances and infield systems was estimated at 12.4 km^3, about half the total water used by the country. Most of the losses occurred by seepage, leading to prolific weed growth in

Table 6.4 Estimated water table depth thresholds at which crop yields begin to be negatively affected. (Adapted from Houk et al., 2006)

Crop	Threshold (cm)
Alfalfa (*Medicago sativa*)	134
Maize (*Zea mays*)	113
Melon (*Cucumis melo*)	100
Sorghum (*Sorghum bicolor*)	96
Wheat (*Triticum aestivum*)	85
Onion (*Allium cepa*)	56

numerous depressions filled with water. A survey of the reclamation status of land under irrigation showed that only 16% of the total land area was classified as good for agricultural use, about 37% reckoned as satisfactory, and about 45% deemed to be unsatisfactory due to the effects of high water table and salinity (O'Hara, 1997).

As at 2001, only 190 million ha of the world's arable land (including rainfed areas) are drained (Table 6.5). The USA with the largest amount of arable land also has the highest amount of drained land.

6.3.2 Soil Salinity

Salt-affected soils classified as saline, sodic, or saline–sodic (Table 6.6) cover over 830 million ha of the earth surface (Table 6.7). About 48% of these are saline soils.

Table 6.5 Five countries with the largest drained areas. (Adapted from Schultz, 2001)

Country	Total Land Area 10^6 ha	Arable Land 10^6 ha	Drained area 10^6 ha
USA	936	188	47
China	960	96	29
Indonesia	190	30	15
India	329	170	13
Canada	997	46	10
World	13,000	1,512	190

Table 6.6 Classification of salt-affected soils

	Electrical conductivity (dS m^{-1})	Exchangeable sodium percentage (%)	pH
Saline	>4	<15	<8.5
Sodic	<4	>15	>8.5
Saline–sodic	>4	>15	<8.5

Table 6.7 Regional distribution of salt affected soils. (Based on FAO dataset available at http://www.fao.org/ag/agl/agll/spush/table1.htm. Saline–sodic soils are usually included with sodic soils because both have similar properties)

Regions	Total area 10^6 ha	Saline soils 10^6 ha	%	Sodic soils 10^6 ha	%
Africa	1899.1	38.7	2.0	33.5	1.8
Asia/Pacific and Australia	3107.2	195.1	6.3	248.6	8.0
Europe	2010.8	6.7	0.3	72.7	3.6
Latin America	2038.6	60.5	3.0	50.9	2.5
Near East	1801.9	91.5	5.1	14.1	0.8
North America	1923.7	4.6	0.2	14.5	0.8
Total	12781.3	397.1	3.1	434.3	3.4

Table 6.8 Global extent of human-induced salinization. (Adapted from Oldeman et al., 1991)

Continent	Light 10^6 ha	Moderate 10^6 ha	Strong 10^6 ha	Extreme 10^6 ha	Total 10^6 ha
Africa	4.7	7.7	2.4	—	14.8
Asia	26.8	8.5	17.0	0.4	52.7
South America	1.8	0.3	—	—	2.1
North and Central America	0.3	1.5	0.5	—	2.3
Europe	1.0	2.3	0.5	—	3.8
Australia	—	0.5	—	0.4	0.9
Total	34.6	20.8	20.4	0.8	76.6

Soil salinity refers to a situation in which the presence of salts renders the soil sterile, whereas sodicity (also referred to as alkalinity) is caused by the specific effect of sodium ions adsorbed on clay particles. This leads to deflocculation of soil colloids and reduction of soil porosity. Soil salinity results from natural and artificial causes. The natural causes could be due to the influence of climate and geology, whereas anthropogenic causes include saltwater intrusion, application of fertilizers and soil amendments, and irrigation.

Human-induced salinity (secondary salinization) occurs on about 77 million ha (Table 6.8), with about 45 million ha of global irrigated cropland salt-affected. Secondary salinization markedly affects productivity and is one of the causes of desertification in dryland areas.

Soil salinity affects crop growth as a result of increase in the osmotic potential of the soil solution. Plants growing in saline soils expend more energy to extract water from the soil. Different plants exhibit different degrees of salt tolerance. Highly salt-tolerant plants (halophytes) tend to absorb salt from the soil solution and sequester it in the vacuoles, with organic compatible solutes helping to regulate osmotic pressure in the cytoplasm. Most crops are salt sensitive (glycophytes) and tend to exclude sodium and chloride from the shoots, relying more heavily on the synthesis of organic osmolytes. In extreme conditions of very high salinity, the external osmotic potential may be depressed below that of the cell water potential, thus resulting in the net outflow of water from the plant and plasmolysis. The sensitivity of plants to salinity depends on climate, soil fertility, soil physical conditions, and physiological stage of crop development. The salt tolerance of some crops is presented in Table 6.9, whereas the effect of soil salinity on crop yields is presented in Table 6.10.

Due to the increased prevalence of irrigation-induced salinity, increased tolerance of crops is needed to sustain food production. Munns (2005) recently provided a review detailing the major adaptive mechanisms of salt tolerance at the physiological and molecular levels. Strategies to improve plant performance on saline soils in crop-breeding programs are also discussed.

The quality of water applied during irrigation affects soil salinity and sodicity, acidity, nutrient availability, soil structure, and crop yields. The salinity of irrigation

Table 6.9 Salt tolerance for selected crops. (Adapted from Mass, 1993)*

Crop	Parameter on which tolerance is based	Electrical conductivity Threshold (dS m^{-1})	Rating
Cotton *(Gossypium hirsutum)*	Seed cotton yield	7.7	T
Groundnut *(Arachis hypogaea)*	Seed yield	3.2	MS
Rice *(Oryza sativa)*	Grain yield	3.0	S
Rye *(Secale cereale)*	Grain yield	11.4	T
Sorghum *(Sorghum bicolor)*	Grain yield	6.8	MT
Sugarcane (*Saccharum officinarum*)	Shoot dry weight	1.7	MS
Carrot (*Daucus carota*)	Storage root	1.0	S
Wheat (*Triticum aestivum*)	Shoot dry weight	4.5	MT
Strawberry *(Fragaria x ananassa)*	Fruit yield	1.0	S
Tomato *Lycopersicum esculentum*	Fruit yield	2.5	MS

*The electrical conductivity refers to that of the soil water while paddy rice is submerged; for rating, T = tolerant, S = sensitive, MT = moderately tolerant, MS = moderately sensitive.

Table 6.10 Effects of soil salinity on potential yields (%) of selected crops. (Adapted from O'Hara, 1997)

	Soil salinity (dS m^{-1})					
	2	4	6	8	12	16
Winter barley (*Hordeum vulgare*)	100	100	100	100	80	60
Cotton (*Gossypium hirsutum*)	100	100	100	94	71	47
Rice (*Oryza sativa*)	100	88	63	38	0	0
Maize (*Zea mays*)	96	72	48	24	0	0
Pepper (*Capsicum annuum*)	93	65	37	8	0	0
Carrot (*Daucus carota*)	86	58	30	1	0	0
Alfalfa (*Medicago sativa*)	100	86	71	57	29	0

water is the sum total of inorganic ions and molecules, the major components being Ca, Mg, Na, Cl, SO_4, and HCO_3. In order to prevent accumulation of salts up to toxic levels, it is necessary to ensure that inputs do not exceed the rate of removal from the soil. The hazard of salinity caused by irrigation water containing salts depend on soil conditions, climatic conditions, crop species, and the amount and frequency of irrigation applied. The degree of limitation posed by irrigation water quality is presented in Table 6.11. Generally, irrigation water with electrical conductivity below 0.7 dS m^{-1} poses virtually no danger to crops, whereas values above 3.0 dS m^{-1} may markedly restrict crop growth.

One other important measure of irrigation water quality is the sodium adsorption ratio (SAR), defined as

$$SAR = \frac{Na}{\sqrt{\dfrac{Ca + Mg}{2}}}$$

SAR is an indication of the activity of Na in exchange reactions. High alkalinity of irrigation water (manifested at pH above 8.5) indicates the predominance of sodium in the solution, and the potential for soil sodicity.

Table 6.11 Degree of limitation of water quality for crop production. (Modified from FAO, 1985)

	None	Slight	Severe
Salinity (Assessed using electrical conductivity or total dissolved solids)			
Electrical conductivity, EC_w (dS m^{-1})	<0.7	0.7–3.0	>3.0
Total dissolved solids (mg l^{-1})	<450	450–2000	>2000
Infiltration (Assessed using electrical conductivity and sodium adsorption ratio)			
Sodium adsorption ratio (%)	Electrical conductivity, EC_w (dS m^{-1})		
0–3	>0.7	0.7–0.2	<0.2
3–6	>1.2	1.2–0.3	<0.3
6–12	>1.9	1.9–0.5	<0.5
12–20	>2.9	2.9–1.3	<1.3
20–40	>5.0	5.0–2.9	<2.9

6.4 Leaching Requirements and Soil Amendments Under Irrigation

Water is usually applied in an amount greater than evapotranspiration to flush away excess salts in the root zone. However, such application of water should take cognizance of the depth of water table and the lateral rate of flow of groundwater. Otherwise, excess water could result in the lowering of the water table leading to buildup of salts at intervals between irrigations. The application of optimum water quantity to prevent water table rise requires the installation of artificial drainage. The standard leaching requirement is to ensure that the maximum concentration of soil solution in the root zone is below 4 dS m^{-1}. However, several factors should also be considered when determining leaching requirements. This include the spatial and temporal variation in the concentration of the soil solution and the degree of salt tolerance of crops which depends on factors such as temperature, humidity, soil matric potential, nutrient availability, and soil aeration (Hillel, 2000). The inherent spatial variability of soils could reduce the efficiency of the leaching process under flooding. Under uniform flooding, water percolates at each spot at a rate dependent on the local infiltrability of the soil. Sprinkling could be more efficient in the removal of salts if carried out at a nonuniform rate adjusted to the properties of soil in the field. Provided that the application rate is lower than the maximal infiltration capacity of the soil, the flow inside the soil will take place under unsaturated conditions. Under drip irrigation, salts tend to accumulate over the periphery of the wetted volumes of the soil, which could adversely affect crop growth in subsequent years. The use of portable sprinklers as a supplement to remove salts in the topsoil is more useful than using drip irrigation alone. Leaching is generally enhanced in permeable soils. Sandy soils are naturally permeable, whereas, clayey soils are permeable only if they are well aggregated.

The application of soil amendments to reduce swelling and dispersion of soil colloids also enhances the leaching process. Flocculating agents such as Ca and Mg are used to replace exchangeable Na on the colloidal surface. Gypsum (CaSO$_4$·2H$_2$O), the commonly applied soil amendment dissolves in the soil solution either until its solubility is reached or the supply in the soil is exhausted. During evaporation, gypsum precipitates faster than salts of higher solubility. Gypsum requirement for replacing exchangeable sodium at the soil colloidal surface is a function of the initial exchangeable sodium percentage (ESP), the soil cation exchange capacity (CEC), and the depth of soil layer to be treated.

6.5   Groundwater Drainage

Artificial drainage is the removal of excess water within the soil so as to lower the water table or prevent it from rising. Drainage maintains favorable aeration and a net flux of salts away from the root zone. It reduces surface runoff and erosion, and improves soil tilth and machine trafficability. It is generally carried out by means of ditches, pipes, or "mole channels" into which groundwater flows as a result of the hydraulic pressure gradients in the soil. The design of optimal drainage system depends on soil type, topography, the climatic and hydrological regime, and the crops to be grown.

Drains are made to direct the excess water by gravity or by pumping to the drainage outlet. The outlet may be a stream, a lake, or the sea. Where disposal into any of this is not possible, an evaporation pond may be used. The use of such closed drainage basin is fraught with certain environmental problems, including negative effects on animal species that use such ponds as their habitat, and the potential risk that such a pond could become habitat for mosquitoes and other human-disease vectors. Deep well injection, a process in which water is pumped into a well for placement into porous rock below the ground surface is another option for disposing drainage water. In regions with acute water scarcity, the drainage water may be reused for irrigation. This however requires the effluent to be free from salts and trace metals so as not to pose any problems to crop production.

Drainage water could also be mixed with nonsaline irrigation water to increase water supply for irrigation. The quantity of nonsaline water required for the blending depends on crop salt tolerance, salinity of the drainage water, and desired yield potential (Table 6.12).

There is increased interest in the utilization of wetlands in the management of drainage effluents (FAO, 1997). Wetlands are particularly efficient in removing sediments, N and P, through complex interactions among plant, soil, and hydrologic parameters to filter and sequester pollutants. The use of wetland for drainage water management depends on residence time, flow rate, hydraulic roughness, and wetland size and shape (FAO, 1997). The water supply to the wetland must also be sufficient to provide an excess amount to discharge and prevent salt buildup.

Saline drainage effluents may also be used in a system of agroforestry. Salt-tolerant trees have the capacity to thrive when irrigated with saline water, and also have the potential to lower the water table by the extraction and transpiration of water from deeper layers in the soil, thereby reducing the volume and expense of

Table 6.12 Proportion of saline water (%) that can be mixed with nonsa-line irrigation water ($EC_w = 0.8$ dS m^{-1}) to achieve a potential yield of 100% and 80% for selected crops. (Modified from Qadir and Oster, 2004)

	EC_w			
	4	6	8	10
Crop	Saline water proportion (%)			
100% yield				
Lettuce (*Lactuca sativa*)	2	2	1	1
Alfalfa (*Medicago sativa*)	14	9	6	5
Tomato (*Lycopersicon esculentum*)	25	15	11	9
Cotton (*Gossypium hirsutum*)	100	62	44	35
80% yield				
Lettuce (*Lactuca sativa*)	37	23	17	13
Alfalfa (*Medicago sativa*)	80	52	39	31
Tomato (*Lycopersicon esculentum*)	78	48	35	27
Cotton (*Gossypium hirsutum*)	100	100	100	100

drainage needed in an area. Some of the trees that could be used include species of eucalyptus, acacia, poplar, and tamarisk (Lee, 1990).

The economic and environmental implications of conventional drainage have made some researchers, for example, Konucku et al. (2006) to propose the concept of *dry drainage* as an alternative. Dry drainage entails setting aside part of the land (e.g., fallow or abandoned land) as a sink drawing a flux of groundwater and salt from the irrigated area. The suitability of dry drainage for wheat–cotton cropping pattern was investigated by Konucku et al. (2006) using a simulation model. The high points of the simulation results are outlined below:

- Dry drainage satisfied the necessary water and salt balance at an estimated irrigated/fallow area ratio of 1.25 when the water table depth was 150 cm.
- The rate of decrease in cumulative evaporation by the fallow area over a period of 30 years was greater during the first decade, but declined toward the end of the period. This trend primarily reflects the rate of salt removal from the soil.
- A shallow depth (100 cm) increases the rate of evaporation in the fallow area, but leads to excessive salt accumulation, whereas a deep water table (200 cm) does not provide enough water flux to sustain the required water balance. A depth of 150 cm is optimal in terms of surface evaporation and crop production.
- Lastly, the coarser the soil texture, the larger the size of required sink area, and the higher the silt content, the smaller the sink area for dry drainage for the same water table depth.

The major drawback of the system is that the allocation of relatively large area as much as 50% of potentially irrigable land as evaporative sink may not be attractive to farmers given the increasing pressure on land and the need to maximize benefits. Further research is required in the identification of factors that could help decrease the size of sink area for dry drainage.

6.6 Some Case Studies of Problems Associated with Irrigation

6.6.1 Irrigation with Saline Water

One major implication of dwindling water resources is that saline water has to be used to meet irrigation requirements in some parts of the world. Tedeschi and Dell'Aquilla (2005) reported an irrigation experiment to evaluate the effects of irrigation with saline water on ECe, soil sodicity (ESP), and aggregate stability in a clay–silty soil in Vitulazio, Italy over a 7-year period from 1995 to 2001. Three saline concentrations of irrigation water (0.25%, 0.5%, and 1% NaCl) and two irrigation levels (100% and 40% restitution of evaporation) were applied in a randomized block design. The results indicated an increase in ECe up to a depth of 155–175 cm for all treatments except control. Lower values of ECe at 20 cm depth were explained by leaching due to autumn–spring rainfall that transports the salts down the profile. Beyond 175 cm, ECe generally decreases with depth. ESP also decreases with depth and increases with the quantity of NaCl applied. The 0–40 cm layer of the most saline treatment (1% NaCl under 100% irrigation) took on saline–sodic characteristic with $ECe > 4$ dS m^{-1} and ESP $>15\%$, whereas at 75, 155, and 250 cm depths, the treatments with 0.5% NaCl under 100% irrigation, and 1% NaCl under 100% irrigation took on saline characteristic with $ECe > 4$ dS m^{-1} and ESP $< 15\%$. In all these treatments the pH was lower than 8.5. Study of changes over time revealed that, there were no significant changes in ESP at 20, 75, and 155 cm depths for the control treatment, whereas there was a significant increase in ESP over time (P < 0.01) for the 0.5% and 1% NaCl treatments at these depths. Only the treatment with 1% NaCl experienced an increase in ECe over time at the 20 cm depth at the rate of 0.5 dS m^{-1} year^{-1} (Tedeschi and Dell'Aquilla, 2005). A progressive and significant increase in salinity was observed at 75 cm depth especially as the salt load increases. However, the rate of salt accumulation did not change over time for a given amount of NaCl applied yearly, even though it increased with the amount of NaCl applied. An inverse relationship was observed between the index of aggregate stability in water and NaCl and ESP (P < 0.01), suggesting the degradation of soil structure with an increase in quantity of NaCl.

6.6.2 Land Use Effects on Salt Mobilization

Macaulay and Mullen (2007) employed groundwater modeling to assess potential saltwater movement under various land uses. A flow-tube model in combination with soil-salinity maps derived from airborne electromagnetic survey was used for the study. Five different land-use/management scenarios were simulated in 5-year steps over a period of 100 years. The scenarios consisted of leakage from large water storage characteristic of the local irrigation system at two locations with different horizontal hydraulic conductivity (0.081 and 0.013 m day^{-1}), long-term irrigation of cotton with a deep drainage rate of 77 mm year⁻¹, extensive grazing assuming a groundwater recharge rate of 0.6 mm year⁻¹, and grazing and rainfed wheat with deep drainage of 6 mm year⁻¹ for wheat. The results indicated that in the land-use/management scenario with relatively transmissive regolith and soil (hydraulic conductivity of 0.081 m day⁻¹), groundwater is predicted to rise to the ground surface in 20 years, with a significant increase (> 50,000 m³) in the volume of salt in the saturated zone. No significant risk in lateral salt movement was however observed. In the land-use/management scenario with hydraulic conductivity of 0.013 m day⁻¹, predicted increase in the volume of groundwater was only 13, 700 m³. The leakage of water and the lateral movement of salt onto the floodplain are restricted by the low permeability of the regolith. The long-term irrigated-cotton scenario predicted groundwater reaching the surface within 15 years. However, owing to the presence of an area of low permeability downgradient, increase in subsurface salt transport will be minimal. No risk of waterlogging or dryland salinity was observed for the extensive-grazing scenario. The last scenario consisting of grazing and rainfed wheat was predicted to generate an increase in groundwater (> 67,000 m³) in the area under wheat, with potential salt mobilization restricted to the area under wheat. Modeled water levels are predicted to remain below the ground surface after 100 years, and dryland salinity is not likely to develop.

6.6.3   Combined Effects of Water Logging and Salinity in Colorado

Houk et al. (2006) recently estimated the combined effects of irrigation-induced water-logging and salinization in a 160,000 ha farmland along the Lower Arkansas River of Colorado between 1999 and 2001 cropping seasons. A hydrologic model that generates information on soil salinity levels and water table depth was linked to an economic model that evaluates productivity losses in the area. The lowest average profit of about $390 ha⁻¹ was estimated for 1999, the growing season with the highest average soil salinity levels and shallowest average water table depths. By comparing baseline profits with expected profits without yield reductions associated with waterlogging and salinity, total profits foregone as a result of waterlogging and salinity ranged between $3.1 million in 2000 to $5.4 million in 1999. Houk et al. (2006) further estimated a potential increase of average profits for the farmland area by 39% if salinity and water-logging problems could be ameliorated. Relationship between further soil degradation and agricultural profitability is not linear, suggesting that additional degradation from current level could result in proportionally higher yield losses that could threaten the viability of agriculture along the Lower Arkansas River.

6.6.4   Fertigation and Soil Salinity in Lebanon

Darwish et al. (2005) assessed the integrated impact of fertilization and irrigation on secondary salinity in two areas in Lebanon. The first area is the Bekaa valley characterized by low annual rainfall of less than 250 mm and annual potential evaporation of about 1,600 mm. Soils of the Bekaa valley are low in organic matter (>0.7%), have massive structure with texture varying from loam to silty loam; human impact has led to the removal of the shallow petrocalcic layer. Farmers intensively cultivate the mixed soil material in monoculture or rotation cropping pattern using high fertilizer and water inputs, with groundwater as the only source of irrigation (Darwish et al., 2005). Significant buildup in soil salinity was observed within a relatively short period of time (1997–2000). The proportion of nonsaline soils ($ECe < 2$ dS m^{-1}) decreased from about 35% to 16%, saline soils (ECe of 4–8 dS m^{-1}) increased from 32% to 39%, and highly saline soils ($ECe > 8$ dS m^{-1}) increased from 10% to 15%.

In northeast Lebanon, Darwish et al. (2005) observed no salinity hazards when surface irrigation or sprinklers were used in monoculture (Table 6.13). The lowest salinity value ($ECe = 0.7$ dS m^{-1}) was obtained under rainfed flooding, whereas the highest salinity value ($ECe = 8.6$ dS m^{-1}) was observed under monoculture with drip irrigation system. Salinity generally increased with number of years of farming, and the layer of salt accumulation is shallower for drip irrigation systems (Table 6.13). Poor drainage control amplified the negative response of crops to salinity. The significant alteration of soil resilience due to salt accumulation, inadequate skills for irrigation management, and the absence of effective extension services led farmers to abandon their plots (Darwish et al., 2005).

Darwish et al. (2005) also observed surplus N applications relative to crop removal of 51 kg ha^{-1} for melon, 65 kg ha^{-1} for watermelon, and 319 kg ha^{-1} for tomato. The fertilizers applied were NH_4NO_3 and KNO_3 with partial salinity index of 2.9 and 5.3 respectively.[1] The observed ranges of salinity at 20 cm depth under the crops were 0.3–0.9 dS m^{-1} for watermelon, 2.0–4.2 dS m^{-1} for melon, and 3.5–13.7 dS m^{-1} for tomato. Excessive fertilization is also attributed as one of the major probable sources

Table 6.13 The impact of cropping pattern and irrigation technique on soil salinity in northeast Lebanon. (Adapted from Darwish et al., 2005)

Cropping pattern	Monoculture				Cereal-based rotation
Soil salinity (dS m-1)	<1	1–2	4–6	>6	<1
Irrigation technique	Rainfed flooding	Sprinklers	Drip	Drip	Drip followed by sprinkler
Farming period (years)	1	3	5	7–10	8
Layer of salt accumulation (cm)	No salinity hazard	40–80	0–40	0–40	No salinity hazard

[1] An important criterion of fertilizer management under irrigation is fertilizer salt index. Fertilizer salt index is a measure of the amount of salt induced in the soil solution as a result of the application of a unit of plant nutrients. It is calculated as the ratio of the increase in osmotic pressure produced by the fertilizer relative that produced by the same weight of Na NO$_3$, based on a relative value of 100 (*Western Fertilizers Handbook*, 1995). Another related measure of salt induced by fertilizers is the salt content per kg or element applied, referred to as partial salt/salinity index. The application of fertilizers with high salt index increases the potential for salt buildup, especially when large amounts of fertilizer is applied and recovery of applied nutrients is low.

of soil salinity in greenhouse production in the coastal area of Lebanon, as water quality explained only about 29% of the variation in soil salinity. Owing to poor fertilizer management and irrigation scheduling, soil salinity inside the greenhouses increased 6.5-fold compared to outside soils (Darwish et al., 2005).

6.6.5   Lessons from the Case Studies

The above-mentioned case studies generally illustrate the importance of having an early warning system to detect the onset of problems in irrigated agriculture. The a priori installation of drainage is the prime preventive measure against waterlogging and soil salinity, and when it is not provided, progressive salinization could lead to land abandonment. While irrigation projects can often function without hindrance for years in areas where the land is naturally well drained, late installation of drainage especially in river valleys often results in a series of cascading events that could eventually lead to the collapse of the irrigation projects. Soil salinity is normally monitored by a combination of soil sampling, soil solution sampling, and various in situ devices that measure salinity. Systematic diagnostic testing conducted repeatedly can provide timely warning of the onset of salinity before it becomes severe. Another method to detect early occurrence of salinity is to place salt-sensitive plants at regularly spaced intervals in the irrigation area. The plants may exhibit symptoms of physiological stress that would allow farmers to respond quickly to salt problems. The use of surface-monitoring equipments to detect and map soil salinity (Rhoades et al., 1997) could also be useful.

Depths of the water table can be monitored by means of observation wells and piezometers inserted into the soil to a depth well below the water table. An observation well is perforated to permit free inflow of groundwater along the length of the tube below the water table, whereas a piezometer is an unperforated tube with its only opening at the bottom. As such, a piezometer indicates the hydraulic head (or pressure) of the water at the bottom of the tube, rather than the position of the water table. A set of several piezometers, inserted side by side to different depths, can indicate the vertical gradient of the hydraulic head below the water table. The direction and magnitude of the gradient indicates the tendency of the groundwater, and therefore of the water table to rise or fall.

6.7   Irrigation and Climate Change

Future agricultural water requirements would be primarily determined by the food demand of the growing population. Socioeconomic development would also play a key role in water availability by increasing the competition between agriculture and other sectors for water resources. On the other hand, increases in greenhouse gas emissions associated with socioeconomic development could significantly alter the Earth's climate with the potential to intensify the entire hydrological cycle. The potential

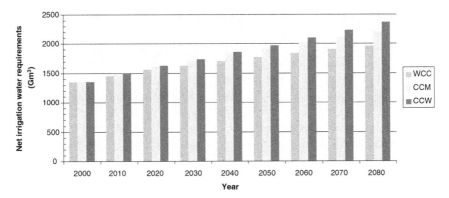

Fig. 6.1 Predicted global net irrigation water requirements without climate change (WCC), with Hadley climate change without mitigation (CCW), and with mitigated Hadley climate change (CCM). (Adapted from Fisher et al. (2006))

impact of climate change on irrigation water requirements has been the focus of some research in the past few years (e.g., Doll, 2002; Rosenzweig et al., 2004). Fischer et al. (2006) recently evaluated the potential changes in irrigation water requirements within a modified Intercontinental Panel on Climate Change (IPCC) socioeconomic scenario. As a result of projected irrigated cropland of 393 million ha by 2080, net irrigation water requirements is expected to reach 1961 Gm^3. Compared to 2000, the additional irrigated water requirement represents an increase of 50% in developing countries and 16% in developed countries. Climate change has an appreciable influence on predicted global net irrigation water requirements: an increase of 409 Gm^3 by 2080 (Fig. 6.1). In fact, 65% of this increase is from higher daily crop water use as a result of warmer temperatures and altered rainfall regime, whereas 35% results from extended crop calendars (Fischer et al., 2006). The impact of climate change will however differ for different parts of the world. In the Mediterranean regions, the combination of low rainfall and high temperature could lead to high evapotranspiration losses and exacerbate water stress for crops. The reduction in availability of good quality water for irrigation would increase the use of saline water, thereby accentuating salinity problems in the region. This implies that crops will have to grow in hotter, drier, and more saline environments (Chartzoulakis and Psarras, 2005). Climate change mitigation is predicted to reduce net irrigation water requirements from 2,370 Gm^3 to 2,212 Gm^3, corresponding to about 39% of water requirements due to climate change. The global annual costs of additional irrigation water withdrawals were estimated at $27 billion, with 63% of these costs anticipated for developing countries (Fischer et al., 2006).

6.8 Conclusions: Making Irrigation Sustainable

The sustainability of irrigation is affected by a myriad of interrelated factors including competition among different sectors for limited water supplies. This calls for the adjustment of water-use patterns for all sectors, including agriculture. As water is becoming increasingly scarce, there is the need for sound and timely guidance on optimizing water quantity and scheduling for irrigation. Economic incentives that encourage water conservation are also required. Other market factors that affect irrigation viability are changing prices of such essential inputs as seeds, fertilizers, pesticides, labor, equipment, and other services, as well as energy. One method to adapt to the changing market conditions is to alter the crop mix in favor of high-valued fruits, vegetables, and industrial crops (e.g., biofuels). Another strategy is to employ more sophisticated technology and management at the field level. Greater emphasis also needs to be given to efficient water and energy use that minimizes environment degradation.

There is also the need to intensify basic and applied research in biotechnology to produce crops requiring less water and nutrients, and increase tolerance to salinity whilst yielding products of superior quality. However, the genetic manipulation of crops also poses risks to natural ecosystems and must therefore be pursued with great caution. In due course, the ability of irrigated agriculture to meet future challenges will largely depend on the progress of new research to enhance adaptation to changing climatic and socioeconomic conditions. Such research is itself costly in the short run, but is a wise investment for private, public, national, and international agencies in the long run.

References

Chartzoulakis, K., & Psarras, G. (2005). Global change effects on crop photosynthesis and production in Mediterranean: The case of Crete, Greece. *Agriculture, Ecosystems & Environment, 106*(2–3), 147–157.

Darwish, T., Atallah, T., El Moujabber, M., & Khatib, N. (2005). Salinity evolution and crop response to secondary soil salinity in two agro-climatic zones in Lebanon. *Agricultural Water Management, 78*(1–2), 152–164.

DFID (Department for International Development). (2000). *Addressing the water crisis: Healthier and more productive lives for poor people.* Consultation Document. London: DFID.

Doll, P. (2002). Impact of climate change and variability on irrigation requirements: A global perspective. *Climatic Change, 54*, 269–293.

FAO (Food and Agriculture Organization). (1985). *Water quality for agriculture.* Irrigation and Drainage Paper 29, Rome: FAO.

FAO (Food and Agriculture Organization). (1997). *Management of agricultural drainage water quality.* Water Reports No 13, Rome: FAO.

Fischer, G., Tubiello, F. N., van Velthuizen, H., & Wiberg, D. A. (2006). Climate change impacts on irrigation water requirements: Effects of mitigation, 1990–2080. *Technological Forecasting and Social Change*, doi:10.1016/j.techfore.2006.05.021.

Hillel, D. (2000). *Salinity management for sustainable irrigation: Integrating science, environment, and economics.* Washington DC: The World Bank.

Houk, E., Frasier, M., & Schuck, E. (2006). The agricultural impacts of irrigation induced waterlogging and soil salinity in the Arkansas Basin. *Agricultural Water Management, 85*(1–2), 175–183.

Konucku, F., Gowing, J. W., & Rose, D. A. (2006). Dry drainage: A sustainable solution to waterlogging and salinity problems in irrigation areas?. *Agricultural Water Management, 83*, 1–12.

Lee, E. W. (1990). Drainage water treatment and disposal options. In *Agricultural Salinity Assessment and Management*. Reston, VA: American Society of Civil Engineers.

Macaulay, S., & Mullen, I. (2007). Predicting salinity impacts of land-use change: Groundwater modelling with airborne electromagnetics and field data, SE Queensland, Australia. *International Journal of Applied Earth Observation and Geoinformation, 9*(2), 124–129.

Maas, E.V. (1993). Testing crops for salinity tolerance. Proceedings of Workshop on Adaptation of Plants to Soil Stresses. In J. W. Maranville, B. V. Baligar, R. R. Duncan & J. M. Yohe (Eds.), *INTSORMIL*. Pub. No. 94-2, August 1–4, 1993. Lincoln, NE: University of Nebraska.

Munns, R. (2005). Genes and salt tolerance: Bringing them together. *New Phytologist, 167*, 645–663.

Nilsson, C., Reidy, C. A., Dynesius, M., & Ravenga, C. (2005). Fragmentation and flow regulation of the world's large rivers systems. *Science, 308*, 405–407.

O'Hara, S. L. (1997). Irrigation and land degradation: Implications for agriculture in Turkmenistan, central Asia. *Journal of Arid Environments, 37*, 165–179.

Oldeman, L. R., Hakkeling, R. T. A., & Sombroek, W. G. (1991). *World map of the status of human induced soil degradation: An explanatory note* (34 pp.). Wageningen, The Netherlands: International Soil Reference and Information Centre.

Qadir, M., & Oster, J. D. (2004). Crop and irrigation management strategies for saline–sodic soils and waters aimed at environmentally sustainable agriculture. *Science of the Total Environment, 323*, 1–19.

Rhoades, JD., Lesch, S.M., LeMert, R.D., Alves, W.J. (1997). Assessing irrigation/drainage/salinity management using spatially referenced salinity measurements. *Agricultural Water Management, 35*, 147–165.

Rosegrant, M. W., & Hazell, P. B. R., (1999). *Rural Asia transformed: The quiet revolution*. Washington, DC: International Food Policy Research Institute.

Rosenzweig, C., Strzepek, K. M., Major, D. C., Iglesias, A., Yates, D. N., McCluskey, A., & Hillel, D. (2004). Water resources for agriculture in a changing climate: International case studies. *Global Environmental Change Part A, 14*(4), 345–360.

Schultz, B. (2001). Irrigation, drainage and flood protection in a rapidly changing world. *Irrigation and Drainage, 50*(4), 261–277.

Tedeschi, A., & Dell'Aquilla, R. (2005). Effects of irrigation with saline waters, at different concentrations, on soil physical and chemical charecteristics. *Agricultural Water Management, 77*(1–3), 308–322.

Vorosmarty, C. J., & Sahagian, D. (2000). Anthropogenic disturbance of the terrestrial water cycle. *BioScience, 50*, 753–765.

Western Fertilizer Handbook. (1995). Produced by the Soil Improvement Committee of the California Fertilizer Association. Sacramento, CA: Interstate Publishers.

World Bank. (2001). *The World Bank and water*. Washington, DC: Water Issue Brief.

World Commission on Dams. (2000). *Dams and development. A new framework for decision-making*. The report of the World Commission on Dams. London: Earthscan.

Chapter 7
Nutrient and Virtual Water Flows in Traded Agricultural Commodities

Ulrike Grote, Eric T. Craswell, and Paul L.G. Vlek

Abstract Globalization and increasing population pressure on food demand and land and water resources have stimulated interest in nutrient and virtual water flows at the international level. West Asia/North Africa (WANA), Southeast Asia, and sub-Saharan Africa are net importers not only of nitrogen, phosphorus, and potassium (NPK) but also of virtual water in agricultural commodities. Nevertheless, the widely recognized declines in soil fertility and problems related to water shortage continue to increase, especially in sub-Saharan Africa. The nutrients imported are commonly concentrated in the cities, creating waste disposal problems rather than alleviating deficiencies in rural soils. And also the water shortage problems continue to contribute to intensified desertification processes, which again lead to increased urbanization and thus water shortage problems in cities. Countries with a net loss of NPK and virtual water in agricultural commodities are the major food exporting countries—the USA, Australia, and some Latin American countries. Understanding the manifold factors determining the nutrient and water flows is essential. Only then can solutions be found which ensure a sustainable use of nutrients and water resources. The chapter ends by stressing the need for factoring environmental costs into the debate on nutrient and water management, and advocates more transdisciplinary research on these important problems.

Keywords Nutrient flows, virtual water flows, international trade, environmental degradation

7.1 Introduction

Vast quantities of water and nutrients are employed in the production of the food that is traded globally. The term "virtual water" refers to the water volume which is used in the process of producing food (Allan, 1997). For example, it has been found that it takes 500–4,000 l of water to grow 1 kg of wheat depending on the location and the technology. In comparison, the production of 1 kg of beef requires about 10,000 l due to the amount of feed consumed by the animals (de Fraiture & Molden, 2004).

A.K. Braimoh and P.L.G. Vlek (eds.), *Land Use and Soil Resources.* 121
© Springer Science+Business Media B.V. 2008

Similarly, it requires different amounts of nutrients to produce 1 kg of wheat, rice, potatoes, or beef, although the regional nutrient requirements vary less based on climate, location, and technologies than compared with water. However, in the case of nutrients, the international transfer in traded commodities is of real not "virtual" nutrients.

Changes wrought by humans in nutrient cycling and budgets are complex and vary widely in magnitude across the globe. In the 35 years between 1961 and 1996, nitrogen fertilization increased 6.87-fold and phosphorus fertilizer use increased 3.48-fold as food production increased 1.97-fold, according to Lambin et al. (2003). On the other hand, Vlek et al. (1997) estimate that 230 Tg[1] of plant nutrients are removed yearly from agricultural soils, whereas global fertilizer consumption of N, P_2O_5, and K_2O is 130 Tg. In the case of nitrogen the estimated 90 Tg from biological fixation must be added to the nitrogen supply. Developing countries now consume half the global fertilizer production, but the use is uneven since cereal crops grown on the irrigated lands of Asia and cash crops receive most of the nutrients. At the other end of the scale, rainfed areas producing subsistence food crops in the tropics, particularly in sub-Saharan Africa, receive little or no fertilizer. In these areas, the farms of poor smallholders develop negative nutrient balances that render continued crop production unsustainable (Stoorvogel & Smaling, 1990). This exploitation of native soil fertility is coupled with the decomposition and decline in soil organic matter that contributes carbon dioxide to climate change.

At the global scale, Miwa (1992) analyzed trends in international trade in food commodities that led to significant negative balances in exporting countries and accumulations in importing countries. Japanese scientists recognized the importance of this problem in their own country which, as a major food and feed grain importer, faces serious nutrient disposal problems due to pollution and eutrophication (Miwa, 1992). Penning de Vries (2006) took this analysis further, describing the environmental problems at both the sources and the sinks of nutrients that move in food commodities. The increasing demand for livestock products in developing countries and the expansion of feed grain use exacerbate the problems (Bouwman & Booij, 1998). Livestock production significantly affects the environment because, as indicated by data for nitrogen, the average efficiency of nutrient conversion from feed to animal products is only 10%, while on efficient dairy farms the range is 15–25% (van der Hoek, 1998). Not only the quality of water is negatively affected through nutrients or antibiotics used in intensive livestock production, but also the total amount of water use as such is immensely high having respective consequences for virtual water contents (Naylor et al., 2005).

The environmental impacts of inter- and intra-national nutrient flows through trade commonly concentrate in the burgeoning cities (Penning de Vries, 2006). For

---

[1] This chapter utilizes SI units as follows: Mg = 1,000 kg (1 metric ton); Gg = ,000,000 kg (1000 t); Tg = 1,000,000,000 kg (or 1 million t); unless specified as the oxide forms P_2O_5 or K_2O, amounts of P and K are converted to, and expressed as uncombined elements; the exceptions are where quoted papers expressed combined NPK data in oxide forms that could not be converted to elemental forms.

example, Faerge et al. (2001) estimated that 20,000 Mg of nutrients were annually imported in food into Bangkok, and that large amounts of nutrients were lost, mainly to the waterways. Coping with large inputs of nutrients in the environment is a major problem facing urban administrations, and the problems are likely to get worse as urban populations grow. Similar problems occur in intensive animal production systems. Nutrients can be recycled through the application of solid or liquid wastes to urban and peri-urban crops and forages but, in spite of the obvious benefits, the extent of recycling is limited in most cities; the potential health problems are a constraint, but safe methods of using wastes are available (Keraita et al., 2003).

Nutrient outflows from fertilized agricultural lands into coastal zones are a widespread problem, particularly in industrialized and developing countries where urea is used excessively (Glibert et al., 2006). In marginal uplands soil erosion by water (and in some cases by wind) also enriches surface waters with nutrients. Some sediment may be deposited and enrich lowland areas. However, annual net ocean outflows of sediments in Asia are as high as 7,500 Tg, representing a major loss of nutrients to the countries concerned (Milliman & Meade, 1983; Craswell, 2000; Syvitski et al., 2005). Global river flows of dissolved inorganic nitrogen to the oceans have been estimated at 48 Tg per annum, whereas 11 Tg per annum is transported to drylands and inland waters (Boyer et al., 2006). Urban wastewater sources of nitrogen represent 12% of the nitrogen pollution in rivers in the USA, 25% in Europe, and 33% in China (Howarth, 2004). These flows of nutrients are in turn affected by human diversions of surface water, such as dams that collect silt and reduce flows to natural wetlands.

The impact of human development on nutrient fluxes in rivers and the seas represent major perturbations to natural terrestrial and aquatic nutrient cycles. In addition, it is important to consider the effects of fertilizers, the production and transport of food, and land transformation for agriculture on fluxes of nitrous oxide because the gas contributes significantly to the greenhouse effect and ozone depletion. Furthermore, nitrates and other nutrients accumulate in groundwater and can affect human health. Nitrogen has been more extensively studied than the other macronutrients, phosphorus and potassium, possibly because human impacts on the global nitrogen cycle extend to industrial perturbations, and atmospheric as well as terrestrial and aquatic phases (Galloway & Cowling, 2002).

While trade in nutrients is mainly considered a reason for environmental problems, trade in virtual water is partly seen as a solution to environmental problems. Trading virtual water is expected to ease some of the pressure related to water shortage and desertification, since countries buying food also purchase water resources, thereby saving water they would have needed for producing the food domestically. It is, for example, estimated that globally 8% of the total water needed for food production can be saved through international food trade (Oki et al., 2003). But also at a national level, situations can arise where food trade becomes necessary due to water shortage problems, as it happens in the case of China: North China has become a major food producing area for the south of China although it faces severe problems of water scarcity. Nearly 10% of the water used in agriculture in the north is used to produce food for south China.

To compensate north China for this virtual water flow, a South–North Water Transfer Project is currently being implemented. This leads to the paradoxical situation of planning to transfer huge amounts of water from the south to the north, while exporting substantial virtual water from the north to the south (Ma et al., 2006). In many countries, the transfer or import of virtual water may minimize the need to build dams and diversions thereby reducing the negative social and environmental impacts (see Chapter 6).

Our brief review above indicates that a substantial, though fragmented knowledge base is developing on the agricultural, ecological, and environmental aspects of alterations to nutrient and virtual water flows and balances at different scales. Many of the ecological aspects are now better understood, but the economic impacts and implications of these perturbations to nutrient cycling have been relatively neglected. One exception in the area of nutrient balances is the work of Drechsel and Gyiele (1999) who developed a framework for the economic assessment of soil nutrient depletion. They showed that the annual cost of replacing nutrients lost from arable land in countries of sub-Saharan Africa ranges from <1% to as high as 25% of the national agricultural Gross Domestic Product (GDP). They estimated that every farm household member contributes about US$32 to the annual nutrient deficit. Related to the annual and permanent cropland in sub-Saharan Africa, the average costs are about US$ 20 ha^{-1} year^{-1}. While this is the case in poor rural areas, at the other end of the spectrum, billions are spent combating environmental pollution from nutrient accretions in urban areas and in lands used for intensive livestock production in both industrialized and developing countries.

This chapter focuses on trends, causal factors, and policy implications related to nutrient and virtual water flows. In Section 7.2, we first review recent estimates of nutrient flows in internationally traded agricultural commodities in 1997 and International Food Policy Research Institute (IFPRI) projections to 2020, and then some estimates of virtual water flows. Section 7.3 discusses the causal factors for both types of flows. We then consider the implications for current agricultural, trade, and environmental policies for nutrient and water flows associated with international trade in Section 7.4, and summarize our results in Section 7.5.

7.2   Trends in Nutrient and Water Balances

International trade of agricultural commodities leads to international flows of nutrients and virtual water. However, few quantitative studies have been conducted estimating current and future nutrient and virtual water flows. In this chapter, nutrient and water flows at the national and global scales will be reviewed. We begin by first outlining the methods used by Grote et al. (2005) to estimate the respective flows and balances, and then proceed to present illustrative results, first for nutrients and finally for virtual water.

7.2.1 Nutrient Flows in Trade

7.2.1.1 Methods

Grote et al. (2005) estimated nutrient flows by following the study of Miwa (1992) who utilized data from the Food and Agriculture Organization (FAO) on food production and trade in the period 1979–1981, calculating nutrient flows using average nutrient contents of the commodities. They also used the IFPRI IMPACT (International Model for Policy Analysis of Agricultural Commodities and Trade) study "Global Food Projections to 2020" by Rosegrant et al. (2001) to estimate future nutrient flows in traded agricultural commodities, using FAO data from 1997 as a baseline. The data included neither nonfood commodities such as wool and cotton, nor industrial commodities such as rubber. The IFPRI data present average values, taken from a variety of sources, and the nitrogen, phosphorus, and potassium contents of the commodities can be seen in the annex of Grote et al. (2005). Fertilizer data used for comparisons with nutrient flows in net trade were converted from oxide to elemental contents. Note that in trade data, an export is normally expressed as a positive value and imports as negative. In the case of nutrient balances, however, the export becomes a negative and any import is a gain to the country. The signs on the IFPRI data converted to nutrient flows therefore had to be changed to reflect the ecological implications.

7.2.1.2 Results on Nutrient Flows in Agricultural Trade

The net global flows of N, P, and K estimated by Grote et al. (2005) total 4.8 Tg in 1997 and 8.8 Tg in 2020. Aggregated data on net flows of nitrogen, phosphorus, and potassium (NPK) in trade vary widely across regions and countries (Fig. 7.1). The countries and regions showing major gains of NPK through imports of traded commodities are WANA and China. Both show major increases between 1997 and 2020, and this is especially true in China, which, according to the IFPRI projections, will increase NPK imported in agricultural commodities from 0.6 Tg to 2.2 Tg. Following the growing urbanization and income-growth trend in China, these imports will probably go to the cities and surrounding intensive livestock production areas, where major nutrient excesses are already causing serious pollution of groundwater and surface waters (Howarth, 2004). Other countries and regions with moderate levels of NPK in imports are EC15, Japan, Southeast Asia, and sub-Saharan Africa. Miwa (1992) has already commented on the environmental problems created by feed grain imports in Japan. From a neutral balance in 1997, South Asia is predicted to become a net importer of NPK in food (0.6 Tg), whereas the NPK imported in food into sub-Saharan Africa is predicted to increase significantly between 1997 and 2020, as is the case with other developing regions such as Southeast Asia. In the latter region, urban income growth and the concomitant growth in demand for animal protein will drive the imports of NPK in meat and feed grains.

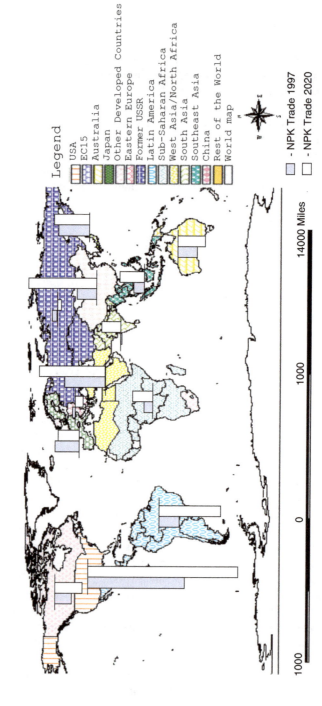

Fig. 7.1 Relative flows of NPK in net agricultural trade in1997 and 2020 (IMPACT model). (Adapted from Grote et al., 2005)

Figure 7.1 shows that the major food exporting countries, the USA, Australia, Latin America, and "other developed regions" (including Canada), show the largest net losses of NPK in agricultural commodity trade. In the case of the USA, the exports of NPK are projected to increase from 3.1 Tg in 1997 to 4.8 Tg in 2020. This represents a major flow of nutrients which has the potential to significantly perturb nutrient cycles in natural ecosystems. The largest relative increase in NPK outflows relate to Latin America which lost 0.65 Tg in 1997 increasing to 1.95 Tg in 2020. Eastern Europe is predicted to increase exports of nutrients to 0.2 Tg, from a low level of import (0.15 Tg) in 1997. The net trade in NPK in the former Soviet Union is generally close to zero in both years.

To assess the relative significance of the NPK flows, Grote et al. (2005) compared the data on nutrient flows in trade (Fig. 7.1) with fertilizer use data in 1997 (for which FAO data were available). The largest exporter of nutrients in trade, the USA, exported only the equivalent of 18% of its fertilizer consumption in 1997. Nevertheless, as discussed in more detail later, the large amounts of nutrients involved do present challenges to US policymakers in the areas of subsidies, trade, and environmental protection. The largest importer of nutrients in absolute terms is China, but because of the large domestic agricultural food production, the amounts of NPK represent only 2% of fertilizer consumption in 1997. Clearly domestic NPK consumption in China dwarfs the NPK imports in food. On the other hand, the trend towards concentration of imported nutrients in the cities may still be a cause of concern since environmental problems are already serious, as noted in the discussion earlier. Japanese imports of the equivalent of 88 to 101% of fertilizer consumption is once again the familiar problem of nutrient overload, as discussed earlier. Although the aggregated data presented for the EC15 do not reveal it, the Netherlands, Belgium, and other small countries that import large amounts of animal feeds have the same environmental problems (Bouwman & Booij, 1998).

In 1997, sub-Saharan Africa imported the equivalent of 26% of fertilizer consumption. At first sight this result may appear to be a positive trend, given that the major problem in sub-Saharan Africa is nutrient depletion in rural areas due to low rates of fertilizer use. However, the data presented are averages across and within countries. If, as discussed above, the nutrients imported in food end up as wastes in the major cities, distant rural lands will not benefit.

Another key issue in sub-Saharan Africa is that while the fertilizer data used by Grote et al. (2005) include nutrients applied to plantation crops, the NPK trade data exclude plantation and industrial crops. Vlek (1993) estimated that in 1987 the export of N, P_2O_5, and K_2O in agricultural commodities, mainly cotton, tobacco, sugar, coffee, cocoa, and tea, was 296 t. In WANA, the import of NPK in food is equivalent to 26% of fertilizer consumption. On the debit side, exports of nutrients from Latin America were 7% of fertilizer use in 1997. While the NPK exports expressed as a proportion of fertilizer use provide a useful estimate of their relative importance, it should be noted that these estimates do not account for the significant biological nitrogen fixation by soybean in the case of Latin America, and by pasture legumes in Australian cropping systems.

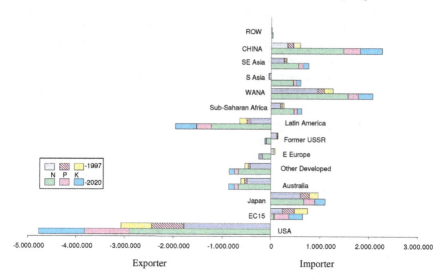

Fig. 7.2 Relative flows of N, P, and K in net agricultural trade in1997 and 2020 (IMPACT model). (Adapted from Grote et al., 2005)

When separating the nutrients N, P, and K in the net trade data (Fig. 7.2), it can be seen that nitrogen is the most dynamic nutrient and after being transformed can move in the atmosphere as well as in aquatic systems. The amounts of N involved in transfers through trade are ecologically significant, especially when the 2020 projections are considered. Potassium transfers are also significant and may contribute opportunities for eventual recycling of this important nutrient, given its high cost of mining and transportation. For phosphorus, the environmental implications for surface water pollution are significant, since it promotes blooms of deleterious blue-green algae that are toxic, and are increasingly afflicting inland water bodies and coastal zones (Wang, 2005).

7.2.2 Virtual Water Flows

7.2.2.1 Methods

Hoekstra and Hung (2005) quantified the volumes of virtual water flows between countries for the period 1995–1999. They multiplied international crop trade flows (ton year^{-1}) by their associated virtual water content (m^3 ton^{-1}). On the one hand, they come to a very conservative estimate of the virtual water trade flows, since they only included the most important crops and did not take crop products like cotton fibers into account. On the other hand, their results are an overestimation as well because they are based on the assumption of optimal growth conditions which hardly exist in reality.

The data used by them were based on the following sources: crop water requirements were taken from FAO's CROPWAT[2] model, the climatic data and crop parameters from FAO's CLIMWAT[3] as well as from Allen et al. (1998), crop yield data from the FAO Statistical Database (FAOSTAT) database, and trade data from the Commodity Trade Statistics Database (COMTRADE) of the United Nations Statistics Division (UNSD).

Oki et al. (2003) and Oki and Kanae (2004) estimated global virtual water trade for the year 2000. They assumed a constant global average crop water requirement throughout the world amounting to 15 mm/day for rice and 4 mm/day for maize, wheat, and barley. The climatic factor, which plays a major role in the water requirement of a crop, was not taken into account as well as the role of the crop coefficient, which is the major limiting factor which determines the evaporation from a crop at different stages of crop growth. The global virtual water flows and the resulting water savings as calculated in these studies are limited to the international trade of only four major crops, namely maize, wheat, rice, and barley.

7.2.2.2 Results on Virtual Water Flows in Agricultural Trade

Oki et al. (2003) and Oki and Kanae (2004) estimated the global sum of virtual water exports on the basis of the virtual water content of the products in the exporting countries (683 Gm³/year) and the global sum of virtual water imports on the basis of the virtual water content of the products in the importing countries (1,138 Gm³/year). This saves 455 Gm³/year as a result of food trade.

The volume of global water savings from the international trade of agricultural products is 352 Gm³/year (average over the period 1997–2001). The largest savings are from international trade of crop products, mainly cereals (222 Gm³/year) and oil crops (68 Gm³/year), owing to the large regional differences in virtual water content of these products and the fact that these products are generally traded from water-efficient to less-water-efficient regions. Since there is smaller variation in the virtual water content of livestock products, the savings by trade of livestock products are less. The export of a product from a water-efficient region (relatively low virtual water content of the product) to a water-inefficient region (relatively high virtual water content of the product) saves water globally. This is the physical point of view. Whether trade of products from water-efficient to water-inefficient countries is beneficial from an economic point of view, depends on a few additional factors, such as the character

---

[2] CROPWAT is a decision support sy stem developed by the Land and Water Development Division of FAO. Calculations of crop water requirements and irrigation requirements are carried out with inputs of climatic and crop data (see http://www.fao.org/AG/agl/aglw/cropwat.stm).

[3] CLIMWAT is a climatic database to be used in combination with the computer program CROPWAT and allows the ready calculation of crop water requirements, irrigation supply, and irrigation scheduling for various crops for a range of climatological stations worldwide (see http://www.fao.org/AG/agl/aglw/climwat.stm).

of the water saving (blue or green water saving[4]), and the differences in productivity with respect to other relevant input factors such as land and labor.

Hoekstra and Hung (2005) find that the global volume of crop-related international virtual water flows between nations was 695 Gm³/year on average over the period 1995–1999. The total water (including irrigation and rainwater) use by crops in the world has been estimated at 5,400 Gm³/year (Rockström & Gordon, 2001), meaning that 13% of the water used for crop production worldwide is not used for domestic consumption but rather "virtually" for export. According to Hoekstra and Hung (2005), for the period 1995–1999, the regions with substantial net virtual water exports were North America, South America, Oceania, and Southeast Asia (Table 7.1). Other net exporting virtual water exporting regions, though less substantial, were the Former Soviet Union, Central America, and Eastern Europe. Regions with significant virtual water import included Central and South Asia, Western Europe, North Africa, and the Middle East. Southern and Central Africa were two additional regions with net virtual water imports, however, to a less significant extent.

Accordingly, the countries with the largest net virtual water export were the USA, Canada, Thailand, Argentina, and India. The largest net import, however, appears to be in Japan, the Netherlands, the Republic of Korea, China, and Indonesia (Table 7.2).

Hoekstra and Hung (2005) find that developed countries generally have a more stable virtual water balance than developing countries. For example in Thailand, India, Vietnam, Guatemala, and Syria, higher fluctuations with peak years in virtual

Table 7.1 Gross and net virtual water exporting and importing regions (1995–1999). (Adapted from Hoekstra and Hung, 2005)

Gross virtual water exporter		Gross virtual water importer		Net exporter (−)/importer (+)	
Region	Gm³ year⁻¹	Region	Gm³ year⁻¹	Region	Gm³ year⁻¹
North America	224	Central and South Asia	196	North America	−206
South America	69	Western Europe	105	South America	−48
Southeast Asia	68	North Africa	51	Oceania	−28
Central America	38	Middle East	41	Southeast Asia	−27
Central and South Asia	30	Southeast Asia	41	Former Soviet Union	−9
Oceania	30	Central America	33	Central America	−5
Western Europe	29	South America	21	Eastern Europe	−1
Former Soviet Union	18	North America	18	Central and South Asia	+166
Eastern Europe	13	Eastern Europe	12	Western Europe	+76
Middle East	11	Former Soviet Union	9	North Africa	+45
North Africa	6	Southern Africa	8	Middle East	+30
Southern Africa	4	Central Africa	3	Southern Africa	+4
Central Africa	1	Oceania	2	Central Africa	+2

---

[4] While green water refers to the productive use of rainfall in crop production with a relatively low opportunity cost, blue water refers to irrigation.

Table 7.2 Top-ten countries exporting and importing virtual water (1995–99). (Adapted from Hoekstra and Hung, 2005)

Country	Net export Volume (10^9 m³/year)	Country	Net import volume (10^9 m³/year)
United States	152	Japan	59
Canada	55	Netherlands	30
Thailand	47	Korea Rep.	23
Argentina	45	China	20
India	32	Indonesia	20
Australia	29	Spain	17
Vietnam	18	Egypt	16
France	18	Germany	14
Guatemala	14	Italy	13
Brazil	9	Belgium	12

water exports were found. As a result it is more difficult to find aggregate figures for developing regions than for developed regions. This fact needs to be factored in when comparing data spaciously and over time.

In addition, Hoekstra and Hung (2005) calculated indicators of national water scarcity, water self-sufficiency, and water dependency. They hypothesized that there is a positive relationship between water scarcity and water dependency because high water scarcity will make it attractive to import virtual water and thus become water dependent. Their hypothesis however was not supported by the data. They explain this by the fact that water scarcity is a driver of international food trade to a limited extent only. Other factors like availability of land, labor, and technology; national food policies; and international trade rules are often more important.

Yang et al. (2003) similarly reviewed whether international virtual water trade is driven by water scarcity. They analyzed the relation between per capita water availability in a country and the net cereal import into the country and found that only below a certain threshold (water availability of about 1,500 m³/year and per capita), the demand for cereal import and thus the virtual water import increases exponentially with decreasing water resources. Above the threshold, there does not seem to be a relationship.

National virtual water balances in 2000 are shown in Fig. 7.3. Countries with net virtual water export are highlighted in blue color while countries with net virtual water imports are highlighted in red.

7.2.3 Comparing Nutrient and Virtual Water Flows

As can be seen from Table 7.3, there are differences with respect to the estimated results on net exports and imports of virtual water between the two studies from Hoekstra and Hung (2005) and Oki et al. (2003). The deviations mainly derive from the high aggregation level, since there is diversity between the countries of one region. In addition, Hoekstra and Hung (2005) use data over a period of 5 years

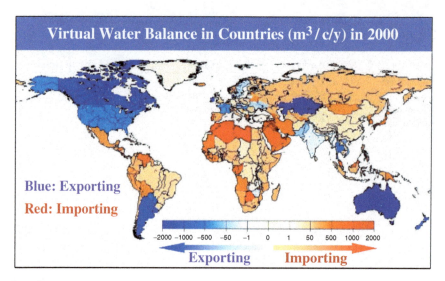

Fig. 7.3 Virtual water balances in 2000. (Adapted from Oki et al., 2003)

Table 7.3 Comparing major estimated results on virtual water and nutrient flows. (Own compilation based on indicated sources)

	Virtual water		Nutrients
	Hoekstra and Hung, 2005	Oki et al., 2003	Grote et al., 2005
Region	Baseline 1995–1999	Baseline 2000	Baseline 1997
North America	–––	–––	–––
Central America	+/–	++	–
South America	+/–	++/––	––
Western Europe	+++/–	++/––	++
Eastern Europe	+/–	+/–	+/–
Former Soviet Union	+	++	+/–
North Africa	++	+++	++
Central Africa	+	++	+
Southern Africa	+	++	++
Middle East	++	+++	+++
Central and South Asia	+++/–	++/––	++
Southeast Asia	++/–	++/––	+++
Oceania	–	–––	–––

Note: a minus equals a net exporter and a plus a net importer; the strength varies between substantial (+++/–––) and less substantial (+/–).

while Oki et al. (2003) base their calculations on only 1 year's data so that certain extreme values in the trade data during that year may weaken the representativeness of the results to some degree.

Comparing the results on virtual water trade with nutrient trade, it can be found that the major regions exporting virtual water are also the major nutrient-exporting countries. The same applies to the imports of virtual water or nutrients, respectively.

7.3   Causal Factors of Nutrient and Water Imbalances

7.3.1   Local-Level Constraints

At the local level, institutional constraints often prevent the adoption of fertilizers in developing countries, resulting in a lack of access to markets, high transaction costs, or often the existence of black markets (Tiessen, 1995). Efficient distribution mechanisms should be created to ensure that fertilizer reaches the farm on time, in adequate amounts, and at minimal cost. An improvement in the marketing of fertilizer is expected to increase its use in developing countries, thus avoiding further soil mining. One strategy to promote fertilizer use is to apply the "mini-pack" method in which small packets of 100 g and 200 g of fertilizers are sold outside shops, in market places, or outside churches (FAO & IFA, 1999). In general, the private sector should be primarily responsible for marketing and distribution of fertilizer, while the government should be responsible for developing and implementing appropriate regulatory and quality control measures to promote more efficient functioning of fertilizer markets. In those areas where markets are underdeveloped, the government may take the lead in developing markets and supporting infrastructure. Furthermore, poor land tenure security, especially in the rainfed mixed farming systems of developing countries, as well as poor access to credit, provide a disincentive for investment in long-term soil fertility improvements (Bumb & Baanante, 1996).

Also with respect to water, institutional failures can have an impact on the water-use efficiency. For example, institutional failures can arise from conflicts within water-user associations or local governments, and from the pricing of water. In addition, poor technology in irrigation can lead to major waste of water.

Finally, in many countries, there is little awareness of efficiency in water and nutrient use because environmental costs are not considered in production decisions, thereby leading to negative externalities.

7.3.2   National Agricultural and Trade Policies

Trends towards nutrient and water imbalances have been triggered by national agricultural policies, promoting a high level of protection in the agricultural sector. In developed countries, like in Western Europe and the USA, agricultural policies have led to a massive use of subsidies resulting in surplus production, export pressures, and financial burdens on the governments. Due to subsidies, the intensity of production has been very high in these countries. The USA, Europe, and Japan spend a total of US$350 billion each year in agricultural subsidies. It is estimated that these subsidies cost poor countries about US$50 billion a year in lost agricultural exports. This is equal to the total aid provided by developed to developing countries (Kristof, 2002). In the USA, for example, subsidies were still expected to increase with the introduction of the US$180 billion farm bill launched in 2002.

Heavily subsidized agricultural sectors tend to distort world agricultural prices and trade patterns thereby limiting developing countries' access to Organization for Economic Development and Cooperation (OECD) markets. As a result, many developing countries have been unable to increase their investments in agriculture, including nutrient inputs leading to soil fertility mining. Due to their limited access to developed countries' markets, their returns have decreased and because of cheap imports from developed countries, their internal markets have also been distorted. Improving developing countries' opportunities to export to developed countries may increase their returns from agriculture, thus allowing them to invest more in their land by increasing their fertilizer input.

In contrast, landlocked and food-deficit countries in Africa are often characterized by low fertilizer use and high fertilizer prices. For these countries, introducing a subsidy may actually help create positive environmental externalities by giving incentives to the farmers to increase their fertilizer use to avoid soil fertility mining.

In contrast, in Australia and New Zealand, the agricultural sector was liberalized and all forms of subsidies were cut in the 1980s. Thus, the market encouraged the farmers to diversify according to their comparative advantage, and not to produce according to the receipt of financial support from the government. In New Zealand, for example, nearly 40% of the average sheep and beef farmer's gross income came from government subsidies in 1984. In 1985, almost all subsidies were removed. About 15 years later, the agricultural sector in New Zealand has grown and is more dynamic than ever. The removal of farm subsidies has proven to be a catalyst for productivity gains. The diversification of land use has been beneficial for the farmers, and the farming of marginal land has declined. Overall, it has been found that the subsidies restricted innovation, diversification, and productivity by corrupting market signals and new ideas. They led to the wasteful use of resources like nutrients and water, negatively impacting on the environment (Frontier Centre for Public Policy, 2002).

In developing countries, input subsidies are commonly used to support farmers. In a survey, FAO found that 68% of developing countries used fertilizer subsidies (FAO & IFA, 1999). In India, for example, the government introduced subsidies for inputs like fertilizer, power, irrigation, and credit to farmers in the late 1960s to foster and encourage agricultural growth. While these subsidies have been critical in getting the "Green Revolution" started, the subsidies are now imposing high costs not only on the environment, but on the fiscal budget. Studies show that fertilizer subsidies, which kept the price of nitrogen low, are now leading to nutrient imbalances lowering crop yield, declining quality of ground water, and the inefficient use and waste of fiscal resources (World Bank, 2001).

National agricultural policies aiming at food self-sufficiency and import substitution policies may also contribute to the inefficient use of nutrients and water at the national level. Instead of importing food products and thus, saving domestic nutrients and/or water, countries often prefer to produce food to feed their own population. This allows them to keep political independence from other exporting countries and insulate them from world market price fluctuations.

Developing countries' past trade policies have often limited the synergistic effect of crops and livestock in nutrient-deficient situations. Imposing high import duties to protect domestic cereal production has pushed cropping into marginal areas and upset the equilibrium between crops and livestock. Also in the former centrally planned economies, industrial systems have greatly benefited from policy distortions and subsidies to feed grain, fuel, and transport which, in many cases, have given these systems a competitive edge over land-based systems.

7.3.3 Global Trends

Despite the slight liberalization success in the agricultural sector in the recent past, international trade volumes in the agricultural sector have increased significantly over time. This growth in trade is partly triggered by the worldwide globalization process. International trade of commodities and products brings along international flows of nutrients and virtual water. The 2020 study by Rosegrant et al. (2001) provides a wealth of insights into major changes between 1997 and 2020 in the supply and demand for major food commodities in developed countries from developing regions. Under their baseline scenario, economic growth, rising incomes, and rapid urbanization in developing countries are driving fundamental changes in the global structure of food demand. As incomes rise, direct consumption of maize and coarse grain will shift to wheat and rice, while the higher demand for meat will increase the demand for feed grain. Agricultural trade will further increase, with wheat leading the cereals and poultry the livestock commodities. WANA, China, and sub-Saharan Africa will increase imports, whereas the USA, Latin America, and Southeast Asia will increase the value of their net exports (Rosegrant et al., 2001).

Urbanization and changing consumption patterns have also been accompanied by a concentration and specialization in livestock production. In areas of high livestock concentrations, excess nitrogen and phosphorus are being produced causing runoffs into groundwater, thereby negatively affecting not only the land but also the water ecosystem around. While in developed countries, stricter environmental regulations push the livestock production increasingly to remote rural areas, in many developing countries, the concentration of animal production takes place in peri-urban areas which benefit from low transportation costs to the city markets and ports, thereby creating major health and pollution hazards to the urban population.

In Europe, about 60% of the total cereals available are used as animal feed, whereas worldwide, it is only 30%. The production of livestock has been very inefficient since it not only requires a lot of water and nutrients, but also results in high greenhouse gas emission rates to air and groundwater. In input-intensive urban farming, the high load of nutrients can be washed into rivers or groundwater, thus contributing to water eutrophication (Kyei-Baffour & Mensah, 1993). The return of nutrients to land-based systems via manure frequently causes problems due to the high water content and high transportation costs. Manure transportation beyond 15 km is often uneconomical, and mineral fertilizers, which are frequently a cheaper and more readily available source of nutrients, reduce demand for nutrients from manure even further, turning the latter into "waste". In Brittany, for example, farmers bought an additional 80–100 kg of N ha^{-1} year $^{-1}$ of inorganic fertilizer, despite the fact that the manure nitrogen would have been sufficient in terms of nutrient intake (Steinfeld et al., 1996). However, it has been also realized that some of these nutrients might reenter the system by contributing to the nutrient requirements of irrigated crops (Cornish et al., 1999). Thus to some extent the off-site costs of water eutrophication might be balanced by the fertilizer value of the water.

7.4   Policy Implications

There are different options as to how nutrient and water imbalances and inefficiencies can be addressed locally, nationally, and globally (Hoekstra & Hung, 2002, 2005).

7.4.1   Local-Level Policies

At the local level, a more efficient natural resource use can be often achieved by creating more awareness among the producers. This can lead to the adoption of diverse voluntary measures and practices which have the ability of minimizing adverse environmental effects. Codes of best practice, voluntary measures like eco-labeling, or the promotion of the "polluter pays principle" (PPP) aim at internalizing environmental costs resulting from nutrient overflow or waste of water at the farm level. For example, a case study in Pennsylvania has showed that the net farm income without natural resource accounting equalled US$80 compared to US$27 when the environmental costs were factored in. Clearly, there is a large gap between private and social costs in the presence of environmental externalities (Faeth et al., 1991; Runge, 1996).

In addition, prices and technologies have been found to play a key role in the efficient use of water and nutrients. Certain price incentives are conveyed by charging prices which take environmental externalities into account, while the use of water- and nutrient-saving technologies can be promoted in the production and processing process. The latter can for example refer to erosion-control measures or minimum-tillage technologies. In addition, it should be mentioned that the establishment of nutrient accounting systems combined with financial accounting systems has been suggested as an important management instrument at the farm level (Breembroek et al., 1996).

Furthermore, there are possibilities to optimize nitrogen and water efficiency by integrating livestock and crop production at the local level. To varying extents, mixed farming systems allow the use of waste products of one activity (crop by-products, manure) as inputs to another activity (as feed or fertilizer). Thus, they are, in principle, beneficial for land quality in terms of maintaining soil fertility, and improving the soil-structure stability, increasing the nutrient-retention capacity as well as the water-holding capacity. This contribution is substantial in economic terms. Approximately 20 Tg or 22% of the total nitrogen fertilization of 94 Tg, and 11 Tg or 38% of phosphate is of animal origin, representing about US$ 1.5 billion worth of commercial fertilizer (FAO, 1997).

7.4.2 Policy Changes at the National Level

At the national level, there are a number of control measures available to regulate the use of nutrients and to reduce their negative effects on soil and water resources. In the European Union for example, the Nitrate Directive suggests a prescribed period for fertilizer application, restrictions of applications on sloping land and near water courses. In addition, there are similar regulations with respect to the use of water (Monteny, 2002).

Improvements towards a more efficient use and allocation of nutrients and water resources can also be made at the national level where governments have to decide on the allocation of water and nutrient resources. They can decide, for example, on how much water should be allocated to which sector (industry, agriculture, etc.), or whether to promote the production of certain basic food products to reach food self-sufficiency, or whether to import commodities to save nutrients and/or water. Accordingly, water-use efficiency can be improved by reallocating water to those purposes with the highest marginal benefits.

As Miwa (1992) points out, each country must consider the consequences of nutrient and virtual water flows in food trade to its own ecosystems. For example, since livestock industries based on imported feeds create massive nutrient-disposal problems, the possibility of importing livestock products directly presents a solution at the national level, though shifting the problems to other countries.

Reductions in fertilizer use may also be achieved by shifting to a human diet with less animal protein. However, human diets are generally not the subject of policy formulation but can be influenced through strategies such as marketing and raising public awareness (van Egmond et al., 2002).

7.4.3 Global-Level Policies

Whereas much research has been done on water-use efficiency at the local and river-basin levels, only little efforts have been made to analyze water-use efficiency at the global level.

7.4.3.1 Trade Liberalization

It is increasingly suggested that international food trade can be used as an active policy instrument to mitigate water scarcity and reduce environmental degradation due to high nutrient concentrations in the soil. A water-scarce country can import products that require a lot of water in their production and export products or services that require less water. Instead of aiming at food self-sufficiency, water-short countries may want to import food from water-rich countries. This is, for example, already happening in Egypt, a highly water-stressed country. In 2000, Egypt imported 8 Tg of grains from the USA, thus saving some 8.5 billion m^3 of irrigation water—one-sixth of the annual releases from the Aswan High Dam. Also Japan, the world's biggest grain importer, would require an additional 30 billion m^3 of irrigation and rainwater to grow the amount of food it imports.

However, saving domestic water resources in countries that have relative water scarcity by the mechanism of virtual water import (import of water-intensive products) also has a number of drawbacks. First, importing food products requires sufficient foreign exchange; second, the importing countries move away from food self-sufficiency which increases their political dependence; third, importing instead of producing domestically results in increased urbanization in importing countries as import reduces employment in the agricultural sector; fourth, there is a reduced access of the poor to food due to decreased agricultural production; and fifth, there is an increased risk of a negative environmental impact in exporting countries, which is generally not accounted for in the price of the imported products. Thus, while enhanced virtual water trade to optimize the use of global water resources can relieve the pressure on water-scarce countries, it may create additional pressure on the countries that produce the water-intensive commodities for export. The potential water saving from global trade is only sustainable if the prices of the export commodities truly reflect the opportunity costs and negative environmental externalities in the exporting countries.

A major improvement in the management of nutrients and water is also expected to result from liberalizing agricultural trade by cutting subsidies and reducing other kinds of policy distortions. In addition, as pointed out by Wonder (1995), input subsidies often mostly benefit those least in need, including large-scale farms and those with higher incomes. They also tend to distort the input mix used for farming through their encouragement of decisions based on support rather than commercial or production criteria. Runge (1996) concludes in his paper on subsidies in the agricultural sector that the elimination of most forms of subsidies in agriculture would produce a double benefit. This double benefit arises from an increasing economic efficiency and also from the reduced indirect negative environmental impacts of artificially expanded production. In Bangladesh, for example, fertilizer subsidies were phased out as part of the overall market liberalization. In China, subsidies were removed by simultaneously increasing producer prices in compensation (FAO & IFA, 1999).

7.4.3.2 Nutrient and Virtual Water Trading Systems

Nutrient and virtual water trading systems have been developed in the recent past to effectively reduce pollution from nonpoint sources like agriculture and urban runoff. Emission trading is being proposed as a new instrument in environmental policy to potentially solve nutrient discharge or water-quality problems. It has been used to create markets in air or water pollution. An overall level of permissible pollution is being set so that the permissible aggregate level of pollution is lower than the current level; thus, an artificial scarcity is created and the permits acquire market value. Producers who wish to expand their production, for example, have the choice to either reduce the pollution from their own production sites or buy permits from others. This way, trading increases flexibility and reduces costs by allowing producers to adapt their own facilities or to compensate others for comparable reductions. Sources with low treatment costs on the other hand may reduce their own effluents beyond legal requirements, generate a credit, and sell these credits to dischargers with higher treatment costs. This flexibility produces a less expensive global outcome while achieving—and even going beyond—the mandated environmental target. The desired reduction in pollution occurs at least cost. Faeth (2000) compared different policy approaches to reduce phosphorus loads in specific watersheds in the USA. He estimates that, based on trading, costs of about US$2.90 per pound of phosphorus removed arise, compared to almost US$24 per pound for conventional point-source requirements. However, there have been also cases where administrative and transaction costs have been high, making trading difficult. King and Kuch (2003) note for example that there are about 37 trading programs in the USA but only a few actually traded nutrient credits. They find that institutional obstacles are significant but can be overcome.

7.5 Summary and Conclusion

The expansion of agricultural production and growing trade in agricultural commodities to meet the needs of an increasing population has transformed vast areas of the land surface and also affected the global water system. As a result, nutrient and water cycles have been perturbed in the soil–water–plant–atmosphere continuum at a range of scales, from local to global levels. The contrasts between the nutrient and water balance in nutrient- and water-deficit and water-surplus countries reflect the large disparities in wealth and agricultural policies between less developed and industrialized countries, respectively. The impact of the international movement of nutrients and virtual water in traded agricultural commodities has been particularly neglected as an area of study. This chapter addresses these research gaps by reviewing and presenting data on global nutrient and virtual water movements and their causes, and by considering government policies that impinge on nutrient and virtual water availability and flows.

Nutrient and virtual water movements in traded agricultural commodities provide a unique international dimension to nutrient and virtual water flows. With respect to nutrients, we obtained net trade data for major food commodities for 1997 from FAO and for 2020 used projections from the IFPRI IMPACT model. Commodity trade data for different regions and countries were converted to weights of N, P, and K using average nutrient contents from the literature. The results show that international net flows of NPK vary widely across regions, but amount to 4.8 Tg in 1997 and are projected to increase to 8.8 Tg in 2020, representing a major human-induced perturbation of global nutrient cycles. Major net importers of NPK and virtual water in traded agricultural commodities are WANA and sub-Saharan Africa. Nevertheless, in sub-Saharan Africa, there is a widely recognized problem of soil nutrient depletion, as nutrients imported in food and feed commodities are commonly concentrated in the cities creating waste disposal problems rather than alleviating deficiencies in rural soils. Also the water shortage problems continue to contribute to intensified desertification processes which are linked to increased urbanization and thus water shortage problems in cities. Countries with a net loss of NPK and virtual water in agricultural commodities are the major food exporting countries—the USA, Australia, Canada, and some countries of Latin America. In Southeast Asia, the picture is more heterogeneous. While some countries within the region are major net importers China, Korea (Rep. of), Indonesia others are major net exporters (Thailand, Vietnam). It is thus difficult to make a general statement for the whole region.

A wide range of policy measures influence agricultural trade, nutrient and virtual water flows, and balances. Agricultural trade liberalization and the reduction of production subsidies are expected to reduce excessive nutrient and water use in nutrient- and water-surplus countries and make inputs more affordable to farmers in nutrient- and water-deficient countries. In an ideal world, this should result in a more efficient global allocation of natural resources, including water and nutrients, and reduced environmental costs, although some level of subsidy to developing country farmers may be justified to introduce them to fertilizer technologies. Policies that encourage diversified production systems should have similar effects by ensuring that animal wastes are not concentrated in areas with no opportunities to recycle nutrients on arable crops. For environmental problems related to poor soil or water quality, innovative policy options such as nutrient and virtual water trading are being examined. For nutrient- and water-deficient countries, institutional strengthening and infrastructure development are valid approaches.

The need for environmental costs to be factored into the debate on nutrient and water management is highlighted in the chapter. Such costs include moving and disposing of waste from millions of tons of nutrients in feed grains used for intensive livestock production. It costs governments billions of dollars to establish and control elaborate environmental regulations to avert water and atmospheric pollution, and it costs society even more when these regulations fail. Because of the increased demand for meat as incomes rise in developing countries, meat-exporting industrialized countries will increase production to meet the demand, increasing the hot spots of nutrient and water pollution in those countries, and the costs will be passed on to their taxpayers. Knowledge of the scope and the long-term costs of these problems should prompt societies and politicians in these countries to support reductions in agricultural subsidies and opening up their markets. The resulting economic development would reduce the need for handouts and probably help stem the out-migration from developing countries. We advocate more transdisciplinary research on these important problems, and solutions such as that proposed above. While hydrologists, soil scientists, or engineers determine what is technologically feasible and set maximum allowances, economists identify the relatively cost-effective options and their distributive effects. The results of the research should better inform the society whose perceptions and values play a major role in the final choice of the products they purchase.

References

Allen, R. G., Pereira, L. S., Raes, D., & Smith, M. (1998). *Crop evapotranspiration: Guidelines for computing crop water requirements.* FAO Irrigation and Drainage Paper 56. Rome, Italy: FAO.

Allan, A. (1997). *Virtual water: A long term solution for water short Middle Eastern economies?* Paper presented at the 1997 British Association Festival of Science, Roger Stevens Lecture Theatre, University of Leeds, Water and Development Session.

Asian Livestock. (2000). *Buffalo meat: The most economical source of protein* (pp. 11–13). Bangkok Thailand: Asian Livestock, FAO Regional Office.

Bouwman, A. F., & Booij, H. (1998). Global use and trade of feedstuffs and consequences for the nitrogen cycle. *Nutrient Cycling in Agroecosystems, 52,* 261–267.

Boyer, E. W., Howarth, R. W., Galloway, J. N., Dentener, F. J., Green, P. A., & Vorosmarty, C. J. (2006). Riverine nitrogen export from the continents to the coasts. In *Global Biogeochemical Cycles, 20,* ID GB1S91.

Breembroek, J. A., Koole, B., Poppe, K. J., & Wossink, G. A. A. (1996). Environmental farm accounting: The case of the Dutch nutrients accounting system. *Agricultural Systems, 51,* 29–40.

Bumb, B. L., & Baanante, C. A. (1996). Policies to promote environmentally sustainable fertilizer use and supply to 2020, IFPRI, Washington DC., 2020 Vision Brief 40, October 1996.

Cornish, G. A., Mensah, E., & Ghesquire, P. (1999). An assessment of surface water quality for irrigation and its implications for human health in the peri-urban zone of Kumasi, Ghana. Report OD/TN 95 September 1999, HR Wallingford, UK.

Craswell, E. T. (2000). Save our soils—Research to promote sustainable land management. In *Food and Environment Tightrope* (pp. 85–95). Proceedings of Seminar, 24 November 1999, at Parliament House, Canberra. Melbourne: Crawford Fund for International Agricultural Research.

De Fraiture, C., & Molden, D. (2004). *Is virtual water trade a solution for water-scarce countries?* Bridges, No.10, ICTSD, Geneva, November.

Drechsel, P., & Gyiele, L. A. (1999). The economic assessment of soil nutrient depletion. Analytical issues for framework development. international board for soil research and management. Issues in Sustainable Land Management No. 7. Bangkok.

Faerge, J., Magid, J., & Penning de Vries, F. W.T. (2001). Urban nutrient balance for Bangkok. *Ecological Modeling, 139,* 63–74.

Faeth, P. (2000). *Fertile ground – nutrient trading's potential to cost-effectively improve water quality.* Washington, DC: World Resources Institute.

Faeth, P., Repetto, R., Kroll, K., Dai, Q., & Helmers, G. (1991). *Paying the farm bill: US agricultural policy and the transition to sustainable agriculture*. Washington, DC: World Resources Institute.

FAO. (1997). *Current world fertilizer situation and outlook 1994/95–2000/2001*. Rome: FAO/ UNIDO/World Bank Working Group on Fertilizers.

FAO and IFA. (1999). *Fertilizer strategies* (rev. ed.). Rome: Food and Agriculture Organization of the United Nations and International Fertilizer Industry Organization.

FAO. (2000): Fertilizer requirements in 2015 and 2030. (Available at ftp://ftp.fao.org/agl/agll/ docs/ barfinal.pdf; accessed on 08 January 2004)

FAO. (2002): Animal feed resources information system. (Available at http://www.fao.org/ DOCREP/ 003/W6928E/w6928e1k.htm; accessed 8 March 2000)

Frontier Centre for Public Policy. (2002). Life after subsidies. The New Zealand farming experience – 16 years later. In *Frontier Backgrounder*. Canada: Frontier Centre for Public Policy.

Galloway, J. N., & Cowling, E. B. (2002). Reactive nitrogen and the world: 200 years of change. *Ambio, 31*, 64–72.

Glibert, P. M., Harrison, J., Cynthia. H., & Seitzinger, S. (2006). Escalating worldwide use of urea: A global change contributing to coastal eutrophication. *Biogeochemistry, 77*, 441–463.

Grote, U., Craswell E. T., & Vlek, P. L. G. (2005). Nutrient flows in international trade: Ecology and policy issues. *Environmental Science and Policy, 8*, 439–451.

Hoekstra, A. Y., & Hung, P. Q. (2005). Globalisation of water resources: International virtual water flows in relation to crop trade. *Global Environmental Change, 15*, 45–56.

Hoekstra, A. Y., & Hung, P. Q. (2002). Virtual water trade, a quantification of virtual water flows between nations in relation to international crop trade. Value of Water Research Report Series No.11. The Netherlands: UNESCO-IHE, Delft.

Howarth, R. W. (2004). Human acceleration of the nitrogen cycle: Drivers consequences, and steps towards solutions. *Water Science and Technology, 49*, 7–13.

Keraita, B., Drechsel, P., & Amoah, P. (2003). Influence of urban wastewater on stream water quality and agriculture in and around Kumasi, Ghana. *Environment and Urbanization, 15*, 171–178.

King, D.M., & Kuch, P. J. (2003). Will nutrient credit trading ever work? An assessment of supply and demand problems and institutional obstacles. In Environmental Law Institute, 33 ELR (pp. 10352–10368). Washington, DC.

Kristof, N. D. (2002). Agricultural absurdity. *International Herald Tribune* (IHT), July 6.

Kyei-Baffour, N., & Mensah, E. (1993). Water pollution potential by agrochemicals. A case study at Akumadan. In *Proceedings of 19th WEDC Conference* (pp. 301–302). September, Accra, Ghana.

Lambin, E. F., Geist, H. J., & Lepers, E. (2003). Dynamics of land use and land cover change in tropical regions. *Annual Review of Environment and Resources, 28*, 205–241.

Ma, J., Hoekstra, A. Y., Wang, H., Chapagain, A. K., & Wang, D. (2006). Virtual versus real water transfers within China. *Philosophical Transactions of the Royal Society of Britain, 361*, 835–842.

Milliman, J. D., & Meade, R. H. (1983). Worldwide delivery of river sediment to the oceans. *Journal of Geology, 91*, 751–762.

Miwa. (1992). Global nutrient flow and degradation of soils and environment. Transactions of the 14th International Congress of Soil Science, Kyoto, Japan, August 1990, V, pp. 271–276.

Monteny, G. J. (2002). The EU nitrates directive: A European approach to combat water pollution from agriculture. In *Optimizing nitrogen management in food and energy production and environmental protection* (pp. 927–935). Proceedings of the 2nd international nitrogen conference on science and policy. The Scientific World (2001).

Naylor, R., Steinfeld, H., Falcon, W., Galloway, J., Smil, V., Bradford, E., Alder, J., & Mooney, H. (2005). Losing the links between livestock and land. *Science, 310*.

Oki, T., Sato, M., Kawamura, A., Miyake, M., Kanae, S., & Musiake, K. (2003). Virtual water trade to Japan and in the world. In A. Y Hokstra (Ed.), *Virtual water trade: Proceedings of the International Expert Meeting On Virtual Water Trade, Value of Water Research Report Series No.12*, Delft, The Netherlands: UNESCO-IHE. Available at http://www.waterfootprint.org/ Reports/Report12.pdf (accessed on 07.07.2006).

Oki, T., & Kanae, S. (2004). Virtual water trade and world water resources. *Water Science & Technology, 49*(7), 203–209.

Penning de Vries, F. T. W. (2006). Large scale fluxes of crop nutrients in food cause environmental problems at the sources and at the sinks. In *Reversing land and water degradation: Trends and bright spot opportunities.* Comprehensive assessment of water management for agriculture. Wallingford, UK: CABI.

Penning de Vries, F. T. W. (2006). Large scale fluxes of crop nutrients in food cause environmental problems at the sources and at the sinks. In *Reversing land and water degradation: Trends and bright spot opportunities.* Wallingford, UK: CABI.

Rockström, J., & Gordon, L. (2001). Assessment of green water flows to sustain major biomes of the world: Implications for future ecohydrological landscape management. *Physics and Chemistry of the Earth (B), 26,* 843–851.

Rosegrant, M. W., Paisner, M. S., Meijer, S., & Witcover, J. (2001). 2020 Global food outlook: Trends, alternatives and choices. Washington, DC: International Food Policy Research Institute.

Runge, F. (1996). *Subsidies and environment. Exploring the linkages* (pp. 139–162). Paris: OECD.

Steinfeld, H., de Haan, C., & Blackburn, H. (1996). Livestock – environment interactions. Issues and options. Rome: FAO. Available at http://www.virtualcentre.org/es/dec/Andes/FAO / Summary/index.htm (accessed 31 June 2006).

Stoorvogel, J. J., & Smaling, E. M. A. (1990). *Assessment of soil nutrient depletion in sub-Saharan Africa: 1983–2000.* Report 28, Wageningen, The Netherlands: The Winand Staring Center.

Syvitski, J. P. M., Vörösmarty, C. J., Kettner, A. J., & Green, P. A. (2005). Impact of humans on the flux of terrestrial sediment to the global coastal ocean. *Science, 308,* 376–380.

Tiessen, H. (1995). Scope 54 – Phosphorus in the global environment – Transfers, cycles and management. (Available at http://www.icsu-scope.org/execsum/scope54.htm; accessed 30 July 2002)

van der Hoek, K. W. (1998). Nitrogen efficiency in global animal production. *Environmental Pollution, 102,* 127–132.

van Egmond, K., Bresser, T., & Bouwman, L. (2002). The European nitrogen case. *Ambio, 31*(2), 72–78.

Vlek, P. L. G. (1993). Strategies for sustaining agriculture in sub-Saharan Africa: The fertilizer technology issue. In *Technologies for agriculture in the tropics* (pp. 265–278). Madison, WI: ASA Special Publication 56.

Vlek, P.L. G., Kühne, R. F., & Denich, M. (1997). Nutrient resources for crop production in the tropics. *Philosophical Transactions of the Royal Society of Britain, 352,* 975–985.

Wang, J. (2005). The ecological engineering of HAB: Prevention, control and mitigation of harmful Algal blooms. *Electronic Journal of Biology, 1,* 27–30.

Wonder, B. (1995). Department of primary industries and energy, Australia: Australia's approach to agricultural reform, Washington, DC.

World Bank, Energy Sector Unit. (2001). India: Power supply to agriculture. Vol. 1, Summary Report No.22171-IN.

Yang, H., Reichert, P., Abbaspour, K. C., & Zehnder A. J. B. (2003). A water resources threshold and its implications for food security. In A. Y. Hoekstra (Ed.), *Virtual water trade: Proceedings of the international expert meeting on virtual water trade. Value of water research report series No.12.* The Netherlands: UNESCO-IHE, Delft.

Chapter 8
The Lesson of Drente's 'Essen'[1]: Soil Nutrient Depletion in sub-Saharan Africa and Management Strategies for Soil Replenishment

Henk Breman, Bidjokazo Fofana, and Abdoulaye Mando

Abstract The term "replenishment" is often misleading, as it suggests that soils are poor through depletion by farmers and that soils should be restored to their original state for agricultural production. This philosophy created awareness of the problems confronted by African farmers. It neglects, however, the heterogeneous redistribution of nutrients that is inherent to agricultural land use. Active and passive transport of organic matter causes centripetal concentration of nutrients around farms and villages and maintains or even improves the soil fertility of crucial fields at the cost of surrounding land. The advice to use fertilizers on bush fields in view of the use of compost and manure on compound fields is like "putting the cart before the horse"; the value–cost ratio of using inorganic fertilizer on compound fields is higher than that on bush fields because of the negative organic matter and nutrient balances in bush fields. The integrated use of inorganic fertilizers and organic forms of manure triggers a positive spiral of improved nutrient-use efficiency and improved soil organic matter status. The increase in value–cost ratio of fertilizer use improves access to fertilizer and other external inputs. Where crop–livestock integration is an important component of agricultural intensification, the centripetal concentration can even turn into the opposite; a centrifugal transport that replenishes (planned or unplanned) the depleted surroundings of farms and villages. Active replenishment of depleted soils is no requirement for agricultural development; intensification can start on village fields where fertility is maintained or improved. However, public investment in soils, focusing on reinforcement of the positive effects of the centripetal concentration of organic matter and nutrients, is recommended; it enables farmers to start fertilizer use where even if the compound fields at present do not allow it.

---

[1] Drente: Dutch province dominated by poor glacial sands. "Essen": 1,000-year-old village fields, on which fertility was improved and maintained through active centripetal concentration of organic matter by man (ash, household wastes, mulch, sods, etc.) and his livestock (manure) at the cost of surrounding rangeland and forest. The process finally led to severe degradation of the latter and to moving dunes. The landscape regenerated and Drente regreened when fertilizer use became economically feasible, livestock numbers could be lowered, and milk and meat instead of manure again became the main product of animal husbandry. See also footnote 2.

A.K. Braimoh and P.L.G. Vlek (eds.), *Land Use and Soil Resources.*
© Springer Science+Business Media B.V. 2008

Keywords Soil fertility depletion and replenishment, nutrient-use efficiency, carrying capacity natural resources, compound and bush fields, agricultural development, sub-Saharan Africa

8.1   Introduction

Two widespread approaches regarding the presentation of agriculture in sub-Saharan Africa (SSA) impede agricultural development policies. Both concern soil nutrient depletion. First, this depletion is expressed in monetary terms, for example, to show that exhaustion of soil resources is a crucial component of farmers' income (Van der Pol, 1992) quantified as the investments required for change (Breman & Sissoko, 1998). Second, it is suggested that the depletion has to be rectified through public investments to enable agricultural intensification and development (Buresh et al., 1997; Breman & Sissoko, 1998; Henao & Baanante, 1999).

These popular descriptions made the world aware of problems faced by African farmers. However, they also hinder effective interventions; the required investments seem immense and discourage donors and national policymakers. Farmers are mistakenly indicated as the cause of their own misery, creating a wrong focus for development efforts. This chapter explains the inadequacy of both descriptions, in a way to refocus efforts for agricultural and rural development. The chapter concentrates on the active and passive redistribution of organic matter and the associated nutrients.[2] Control of this process, limiting the passive component at the benefit of the active one, is of utmost importance for the transition from extensive to intensive agriculture. The chapter first explains briefly why agriculture in SSA is still predominantly in the extensive stage, and why even a very intensive and intelligent use of organic matter alone will not change this. Then, it shows how the redistribution of organic matter can be exploited to trigger intensification through the improved accessibility of inorganic fertilizer. Finally, the chapter describes the conditions that must be fulfilled for successful intensification.

8.2   Naturally Poor Soils Combined with Difficult Climates: A Major Constraint for Agriculture Development

8.2.1   The Case of Sub-Saharan Africa

Africa is the world's most ancient land mass. About 90% of its soils lost most of their nutrients during several years of erosion and leaching; only 10% of the soils are relatively young and still have nutrient rich sediments. The nutrient-impoverished

---

[2] Active redistribution: collecting, transporting, and using materials such as manure, household wastes, straw, and sods to maintain or improve the fertility of certain fields. Passive redistribution: concentration of materials in the surroundings of farms and villages by transport mechanisms having different goals, such as livestock coming to pass the night in kraals, bringing wood and straw for construction, and collecting fruits in the forest for home consumption.

soils produce limited plant biomass; consequently, the soil organic matter content is low. The major European agricultural soils, for example, contain at least twice as much soil organic matter as those in SSA.

"Soils in SSA are not in the first place poor through depletion by farmers, but farmers deplete soils because their soils are poor by nature." This is the foundation of Breman and Debrah's (2003) recommendations for reaching food security in this part of the world, justified by decades of research. They explain how poor resource base caused the "green revolution" to bypass SSA and that a very unfavorable value–cost ratio (VCR) seriously hampers the use of inorganic fertilizer. These points are summarized as follows:

- Extremely poor natural resources cause overpopulation at low absolute population density,[3] leading to high prices for external inputs and low prices for agricultural products at the farm gate.
- The efficiency of inorganic fertilizers is low in the soils.

Inadequate socioeconomic and policy environments worsen the situation. Besides social and political instability, favoring consumers at the cost of producers is common, the development of competitive input and output markets is hindered by corruption, lack of quality control, and other factors. The poor resource base of agriculture can be used only partially as an excuse.

"VCRs above two or even three to four are no guarantee that farmers will start applying fertilizers in their fields. Kelly et al. (2005) state correctly that reducing risk and uncertainty plays an important part in improving fertilizer incentives in SSA" (Meertens, 2005). The latter, presenting the evolution of VCRs in 15 countries of SSA, shows that values of 3 or just above 3 are already rare; 11 out of 17 values (maize, cotton, and rice) from the early 2000s are lower. Only during the early 1980s, a period of heavy subsidies, were the VCR values considerably higher.

Binswanger and Pingali (1988), trying to explain the particularities of African agriculture, argue that land is still too abundant to attribute it a value and price high enough to trigger the required investments for change. Van Keulen and Breman (1992) reacted that in most of Africa, land is not abundant at all; poor soils require extended fallow in time and space. Therefore, "extreme land hunger" describes the situation better. What occurs if the natural resources degrade (quasi) irreversibly before the favorable density for attributing a value and price to land is reached? (Breman, 2000).

---

[3] In the West African Sahel and Soudanian savanna, respectively, 6 and 33 persons per km^{-2} can live from the land in a sustainable way. Overpopulation occurs at population densities above this carrying capacity, and sustainability can only be guaranteed by using external inputs, such as inorganic fertilizers, to increase the carrying capacity of the land. The adoption of agricultural technologies based on external inputs has to occur at a low population density in comparison with such changes elsewhere. For example, in Southeast Africa, it occurred when population densities, the inherent road and market infrastructures and domestic market were 10 times higher.

8.2.2   Sub-Saharan Africa is not an Isolated Case

SSA is not the only part of the world where the growth of a population undermines the future of that same population. Mazoyer and Roudart (1998) present a world-wide overview of the history of agriculture. They focus attention on the variation, potentials, and limitations of natural resources, and the maximum population density related to different exploitation systems. Slicher van Bath (1960), limiting himself to West Europe, compares the agricultural evolution on rich marine and fluvial clay soils with that on poor glacial sands in the Netherlands. Long before the existence of inorganic fertilizers, farmers on clay could produce more than that required to feed their families. This enabled them to develop market-oriented production systems early, to encourage (most of their) children to look for labor opportunities outside agriculture, to invest in their land, and to develop inheritance systems that maintained their properties. The situation was opposite on poor sands—a farmer could barely produce enough to feed his family, and every family member was obliged to help make that happen. Production systems focused on self-sufficiency; upon the death of the farmer, his property was divided among the surviving children. After the fallow system became inadequate in view of population pressure, investments in arable land were limited to the active and passive transport of organic matter and their nutrients. Numerous livestock and humans functioned as the primary and secondary tools, and the amounts of manure, hay, leaves, fruits, wood, and sod brought to farm and village created fields that were sometimes more than 1 m higher than the original level. It was at the expense of range, forest, and wasteland that degraded progressively and became moving sands. Entire villages disappeared, and the population reinforced the army of job seekers in the cities. Entirely different landscapes developed on clay and on sand.

Bieleman (1987) studied the history of agriculture on Dutch sands in detail, but limited himself to one rather homogeneous region, the province of Drente. He shows numerous variations in evolution of which the description above is a rather simple generalization, and insists that private initiatives and market opportunities are crucial explanatory factors of agricultural development. Growing markets triggered intensification and diversification, in particular, at the dawn of industrialization in Europe, in spite of the dominance of production for self-sufficiency and the degradation of (part of) the natural resources. The study underlines that even on the scale of a small Dutch province; the distance to external markets is a factor for differentiation. Tenths of kilometers more could imply centuries of delay in intensification when infrastructure is poorly developed, and carts with animal traction primarily provide transport! Many others factors also stress the importance of the context in its agroecological, geographical, socioeconomic, and political sense. It is illustrative that the relative importance of crops and livestock and the attention they receive from farmers vary with product prices in the Netherlands and in Europe. At the turn of the nineteenth century, increasingly available inorganic fertilizers were mainly used to improve livestock feeding as a result of high dairy prices (Bieleman, 1987).

8.2.3 How to Learn from Others and Elsewhere?

The agricultural development stage in SSA today is where Drente was a century ago: overpopulated at low absolute population density, using crop–livestock integration to try to maintain crop yields, having limited access to inorganic fertilizer as an alternative, and using market opportunities as challenges for change. A crucial question concerns the time still allowed for using concentration of organic resources through livestock and manpower to maintain crop yield and production while the population density is increasing. The crux is the fraction of land occupied by fields, and the carrying capacity for livestock of the other part. Only temporarily, herd growth can satisfy the increasing manure requirements of extending fields. Fields used for crop production increase at the expense of grazing land, and the manure, which is produced, must be distributed over an expanding area. Besides, extreme grazing pressure threatens other land uses, from which wood production is the most important under the described circumstances (Slicher van Bath, 1960; Mazoyer & Roudart, 1998).

It is tempting to derive an answer from parallels between the situation in Drente in the past, and, for example, Burkina Faso and Rwanda today. In all three cases, livestock effectives increase rapidly with population density and the inherent extension of fields. In Drente, with the doubling of the area occupied by fields, from 10% to 20%, animal density more than doubled from about 30–80 tropical livestock units (TLU) km^{-2} (time series 1832–1910; derived from Bieleman, 1987, using livestock weights 1.5 times greater than those of tropical animals). In Burkina Faso with the increase in the area occupied by fields from 5% to 30%, animal density increased from about 10 to 40 TLU km^{-2} (data 1995 per province [de Ridder et al., 2004]). In Rwanda with the increased area occupied by fields from 10% to 40%, animal density increased from about 12 to 33 TLU km^{-2} (data 2002 per province; derived from MINECOFIN, 2003).

In spite of parallels, differences between agroecological and socioeconomic conditions are far too large to obtain even an impression of the respite that the latter countries still have before production systems based on redistribution and increasingly intensive use of locally available organic matter collapsed. Drente had to feed more than 60 inhabitants km^{-2} when fields occupied 20% of the land, Burkina has to feed almost 50 inhabitants km^{-2} with an average of 15% of the land occupied by fields; figures for Rwanda are respectively 320 inhabitants km^{-2} and 33%. Whereas soils in Drente and Burkina Faso are predominantly poor (loamy) sands with limited reserves of organic matter and nutrients, Rwanda has regionally extremely fertile volcanic soils. Average cereal yields in Burkina Faso are about 0.8 t ha^{-1} $year^{-1}$. In Drente at the turn of the nineteenth century, they were 1.3 t, and in Rwanda today the production per hectare and per year exceeds 2 t (two harvests per year). In Drente, 65% of the population depended directly upon agriculture; in Burkina Faso and Rwanda it is about 80% and 85%, respectively. Although few people are employed outside agriculture, the extremely high population density implies that the domestic market for agricultural products in Rwanda is nevertheless much larger than that in Burkina Faso. For example, Cour (2001), Tiffen et al. (1994), and Wiggins (1995) stress the opportunities created by large domestic markets for intensification, in West Africa, Kenya, and SSA as a whole.

Despite the differences in agriculture between Drente, Burkina Faso, and Rwanda, striking parallels exist. In the three cases, livestock is increasingly exploited to maintain the productivity of crops, and in the three cases, it causes erosion that undermines the future of agriculture. Moving sand in Drente in the past was as threatening as moving dunes and sheet erosion in Burkina Faso and eroding hillsides in Rwanda today. In the three regions, cereal yield increases under growing population pressure obtained without external inputs such as inorganic fertilizers are about $7\,kg\,ha^{-1}\,year^{-1}$ (Bieleman, 1987; Breman, 1998), similar to the worldwide situation (Wit, 1986). The three regions have the same two options for exploiting increased animal pressure that goes hand in hand with extension of fields at increasing population pressure: depleting all land that is not used for crops to its limits while undermining the future, or using it for advancing to fertilizer use as described in the next chapters (Breman, 1990).

8.3   Redistribution of Organic Matter and Soil Nutrients

8.3.1   Soil Depletion

Stoorvogel and Smaling (1990) created awareness regarding the phenomenon of soil fertility depletion in Africa, with their evaluation of the national agricultural nutrient balances. They determined the sum of nutrient inputs through fertilization, use of organic residues and manures, atmospheric deposition, and sedimentation. Then they subtracted nutrient losses through erosion, leaching, volatilization, and crop uptake, followed by export from fields at harvest. Ten years later, Henao and Baanante (1999) repeated and confirmed their work, after improving some of the tools. Both studies evaluated nutrient balances at spatial scales that range from small plots to the whole continent. Henao and Baanante (1999) presented their results per country, as average values per hectare of cropland. Fields in 23 African countries annually lose more than $60\,kg\,ha^{-1}\,N + P_2O_5 + K_2O$; those in Rwanda with $136\,kg\,ha^{-1}$ are the highest. Losses in South Africa are negligible; Burkina Faso (with $62\,kg\,ha^{-1}$) is at the bottom end of the countries with extremely high depletion. A group of 14 countries has medium-level soil nutrient depletion (-30 to $-60\,kg\,ha^{-1}\,year^{-1}$); only 7 have low ($<30\,kg\,ha^{-1}\,year^{-1}$) or no depletion. Factors that cause high depletion are, for example, high erodibility of soils, high rainfall, and high population pressure. In particular, manure use and inorganic fertilizer counterbalance losses. The countries with no or low depletion have an average annual fertilizer use of $147\,kg\,ha^{-1}\,N + P_2O_5 + K_2O$. Countries with medium losses use on average $20\,kg\,ha^{-1}$, those with high losses only $13\,kg\,ha^{-1}$.

Factors determining the use of inorganic fertilizers are profitability, favorable agroecological conditions (soil and climate), fertilizer aid by donors, fertilizer subsidies, pan-territorial crop and fertilizer prices, access to input credit and to output markets, and other policy-linked factors such as public investments in irrigation and in soil improvement. Social unrest and lack of political and policy stability seriously hinder fertilizer use (Meertens, 2005; Breman & Debrah, 2003). Also, policies of rich countries can have serious negative impacts on fertilizer use in Africa, for example, dumping of meat by the European Union made high quality fodder production in West Africa impossible, and cotton subsidies in the USA and elsewhere threaten intensive cotton production in Africa (see also Chapter 7).

In a recent paper, Henao and Baanante (in press) compare the depletion rate of 1995–1997 with 2002–2004. Even in this short period, the situation has worsened. One country moved from the group of low depletion to medium depletion; two countries moved from high to medium, but five others were added to the list of countries with high depletion.

8.3.2 Redistribution: A Component of the Process of Depletion of Soil Fertility

Mazzucato and Niemeijer (2000) measured organic matter and nutrient content of soils in two villages in Burkina Faso and compared the results with measures from the past. The two villages had different population pressures as well as past and present land-use pressure. They observed that in the densely populated village, soil fertility decline was not or hardly detectable. This made de Ridder et al. (2004) to review literature so that they could revisit the analysis of agricultural production systems in West Africa of the 1980s. That analysis (e.g., Keulen & Breman, 1990) revealed that the population pressure was already so severe that without the use of inorganic fertilizer, decline in soil fertility of both grazing and arable land would lead to decreasing land productivity and jeopardize food security.

De Ridder et al. (2004, p. 2) conclude that "nutrient budgets show negative trends in stocks, which are probably overestimated because lateral in- and outflows are scale-dependent, difficult to estimate and often ignored." They explain that under farming conditions in SSA, decline in soil fertility of rangeland, and forest can hardly be measured in view of "highly variable soil management in space and in time. However, at coarser scales, gradients in soil fertility are detected as a result of centripetal transport of organic material." In other words, the apparent contradictions between Mazzucato and Niemeijer (2000) and van Keulen and Breman (1990) are explained by the difference in focus. The first authors concentrate on fields while the latter regard the system as a whole. Most soil depletion occurs on rangeland and in forests, not on fields.

8.3.3 Quantifying the Redistribution Phenomenon

Hilhorst and Muchena (2000, p. 2), measuring nutrient flows on cropland as detailed as possible all over SSA, focus attention on the role of farmers' management, for example, by comparing compound fields and outfields. They conclude that: "moving sources of nutrients around creates 'hotspots' of good quality, fertile soil, even where the sum of nutrient balances in all fields may be negative at the farm level." Haileslassie (2005, p. 95) arrives at a similar conclusion for mixed farming systems in Ethiopia:

> The positive partial nutrient balances for the intensively cultivated regions of Ethiopia suggest that farmers are applying sufficient inorganic and organic fertilizer to counterbalance the nutrient losses caused by the removal of harvested products and crop residues. In contrast, the full nutrient balances are negative, indicating that the soil nutrient reserves of the country are decreasing.

Ramisch (2005), calculating plot- and household-level soil nutrient balances from participatory exercises and soil sampling, observes overall annual community-level nutrient balances of -9.2 kg ha^{-1} of N, $+0.8$ kg ha^{-1} of P, and -3.4 kg ha^{-1} of K for extreme southern Mali. While considering both crop and livestock systems, his balances are partial; he considers only fields at a distance of 3.5–4 km from compounds. He indeed observes centripetal concentration of nutrients; for example, the average N input in fields increases from about 40 kg ha^{-1} at 3.5 km from home to 100 kg ha^{-1} for the home compound. A positive N balance is found only for the compound field ($+3.8$ kg ha^{-1} year^{-1}); elsewhere the balance is negative (average -15 kg ha^{-1} year^{-1}). Like Bieleman (1987) in the case of Drente, Ramisch focuses on the strong variation between cases:

> "... coefficients attributable to household behaviors matched or surpassed those attributable to distance and .... System differences in household asset ownership, use and resource allocation behavior suggested that much of the diversity seen in the nutrient balances and soil analyses was due to persistent inter-household inequality and the consequent exchanges of agro-pastoral resources." (Ramisch, 2005, p. 353)

Livestock appears to be a crucial asset; comparing three villages that differ in stocking rate between 7, 17, and 20 TLU km^{-2} (average values dry and rainy season), average annual N balances for fields of -15, -3, and $+20$ kg ha^{-1}, respectively, are observed. Without the frequent manure trade, grazing contracts, and purchases of inorganic fertilizer, involving dozens of kg ha^{-1} N, differences annually should still have been larger.

Active and passive transport of organic matter, containing essential mineral nutrients, causes centripetal concentration of these nutrients around farms, villages, and towns. Thanks to this process, the soil fertility of crucial fields can be maintained or even improved at the cost of range-, forest-, and wasteland. Islands of (relative) fertility appear, even in regions with very poor soils, comparable to humid run-on spots through redistribution of rainwater by runoff in (semi-)arid regions. The comparison is useful; redistribution of water and nutrients creates agricultural opportunities that are absent at homogeneous distribution of limited amounts.

Krul et al. (1982) observed the phenomenon of centripetal organic matter concentration and tried to quantify it. Total N, C, and P were respectively 2.75, 3, and 6.8 times higher in the top 10 cm of the soil of a large Sahelian village than at a distance of 6 km; intermediate figures occurred in between. They estimated the export of P by grazing from rangeland far from villages to be in the order of 100 g ha^{-1} year^{-1}. Their primary interest for P should be emphasized. Since phosphorus is a nutrient that has limited mobility and is not volatile, its centripetal concentration should be higher than that of N, as proven by their observations. It implies that the relative dominance of N-deficiency over P-deficiency, as observed for soils in large parts of West Africa (Breman & Van Reuler, 2002), is accentuated in the redistribution process.[4]

Synthesis of data from three other studies and regions confirms the above-mentioned trend (Breman, 2002). The top 20–30 cm of soil of village fields has a total N, C, and P content of 750, 7,900, and 270 mg kg^{-1}, respectively, which is 1.6, 1.7, and 2.6 times higher than observed for the bush or bush fields. Using available P instead of total P, the concentration for the village field is 50 mg kg^{-1}, even 8 times higher than that for the bush field.[5] In view of the mechanisms involved, grazing and wood production, the land further away than the bush fields will have somewhat lower organic matter and nutrient contents than the bush fields. Krul et al. (1982) showed a decrease of total P from 515 mg kg^{-1} in the village to 153, 95, and 76 mg kg^{-1} at 1, 2, and 6 km distances, respectively. The centripetal concentration can be measured, but as others have already stressed, it is generally very difficult to measure a soil fertility decrease in grazing land because organic matter and its nutrients are harvested from large areas in comparison with the land under crops (de Ridder et al., 2004; Turner, 1995). The supply of P by livestock on compound fields, expressed as fraction of total P in the top 25 cm of the soil, is annually about 2% or more; the average depletion of rangeland is only one-hundredth of it.

The parallel increase of the area occupied by fields and the stocking rate of livestock imply that the supply of manure and the mineral nutrients that it contains per hectarage of cropland is rather constant; whereas, the depletion of grazing land is increasing. The example of P above is based on an estimation using the simple ideal but theoretical case of a concentric distribution of land-use types: compound fields (10% of the entire cropland) surrounded by bush fields, surrounded by grazing

---

[4] Observations such as those of Green and Cole (2006) regarding feedlots in the USA confirm this conclusion: "A 20,000 head capacity feedyard requires at least eight sections of irrigated corn to dispose of the manure based upon its fertilizer value for nitrogen, but over 48 sections of irrigated corn is required to dispose of the manure based upon its fertilizer value for phosphorus. Consequently, applying manure to croplands to meet the nitrogen requirements oversupplies phosphorus."

[5] Samaké (2003, p. 85) observes less difference between the concentration of N and P for extremely poor soils of the Malian Sahel. In the top 15 cm of compound fields, with 280 and 6 mg kg^{-1} of N and P-Bray-I, respectively, the N content appears 2 times higher than that of the bush fields, against 2.6 for P-Bray-I.

land (range, forest, and wasteland). As radius of the outside circle of grazing land 7.5 km has been chosen, 15 km is used as the maximum grazing distance (twice the radius if the herd passes the night at the farm). Grazing is done entirely in the outer circle for which 8 h day^{-1} is available. Two-thirds of the manure produced drops on cropland during rest, rumination, and moving. Food intake and quality are those of a typical herd for integrated crop–livestock systems using animal husbandry to maintain crop production (Ketelaars, 1991). The estimation concerns the maximum amount of manure that can be obtained for transporting nutrients from grazing land to arable fields. The maximum supply is about 640 kg manure (DM) per hectare of cropland at homogeneous distribution. Concentrating all on the compound fields, it becomes 6,400 kg ha^{-1} DM. The figure is the same for 10% of the total acreage under crops and a stocking rate of 10 TLU ha^{-1}, as for 30% land under crops combined with a stocking rate of 30 TLU ha^{-1}.

If he succeeds in collecting all manure that is produced during the stay of the herd in the two inner circles, and if nothing is lost during transport, storage, and handling, a farmer has the choice between applying 640 kg ha^{-1} DM on all his fields or 6,400 kg ha^{-1} on his compound field. However, during storage, the mineralization process starts and organic matter will be lost; this results in an increase in the concentration of mineral nutrients in the composted material. In view of the humidification coefficient of about 0.5, at least one-quarter of the manure on DM base will be lost during the average 6 months that it must be stored; what remains is at maximum 480 kg ha^{-1} for all fields or 4,800 kg ha^{-1} for compound fields. Based on data from Sahelian countries, Duivenbooden (1992) finds average mineral nutrient contents for animal manure of 1.3%, 0.3%, and 1.3% of DM for N, P, and K. In other words, on average about 6, 1.5, and 6 kg ha^{-1} are available for all fields or doses of 10 times more on compound fields only.

As already stressed above, losses of the relatively mobile nutrients, N and K (mainly through leaching, but in case of N, also through volatilization), in the period between grazing and manure application have been much higher than that of P. The P content of fodder from Sahelian rangeland is one-tenth of the N content (Penning de Vries & Djitéye, 1982); in manure it appears to be a quarter (Duivenbooden, 1992). In this context, the term "losses" means "out of control of the farmer." At least part of the lost nutrients will still serve crop growth, if the losses occurred in the field.

8.3.4 Reinforcing the Redistribution Phenomenon

A simple ideal, theoretical case was presented above. A series of studies exist from which the potential role of different variables for reinforcement of redistribution can be derived. Turner (1995) identifies them by using a formula that describes the rangeland: "cropland ratio necessary to support manure-supported continuous cropping." Variables that are emphasized are the production capacity of the land, the fraction of cropland for which fertility is still maintained by fallow, the production of leguminous fodder crops (citing Garin et al., 1990), and the grazing radius. They are positively but not linearly correlated with the acreage of cropland supportable by manure in the place where animals spend the night. This acreage increases also by placing the corrals at the outer edge of the continuously cultivated land instead of in the village.

Without any quantification of their effects, Turner (1995) also mentioned livestock, crop, and soil management. Quak et al. (1998) developed modeling and simulation software that enables the quantifying of the contribution of agricultural practices and management to soil organic matter maintenance and nutrient availability. The modeling incorporated animal production systems (mobile and sedentary, and mixed grazing herd versus fattening at the farm), different cropping systems and crops (rainfed and irrigated; cereals, legumes, food, fodder, and cash crops), and crop residue management (burning, mulching, burying, and composting). It is rather obvious that practices such as sedentary stable feeding, production of high-quality fodder, and the collection and composting of as much manure, crop residues, and household wastes as possible all reinforce the redistribution phenomenon and farmers' control over it.

Collection and composting of manure, crop residues, and household waste is a crucial variable. This is illustrated by the data on which the above average animal manure nutrient contents of Duivenbooden (1992) are based. Using less than 20 different sources, the N content already varies between 0.35% and 2.50%, the P content between 0.11% and 0.41%, and the K content between 0.46% and 4.50% of manure dry matter. Factors that improve the contents by reducing losses include the use of straw to absorb urine and decrease leaching, concrete floors, and anaerobic treatment (pit instead of heap composting; e.g., IFDC, 2002; Lekasi & Kimani, 2005).

A strong and socially rather tricky mechanism for reinforcement of redistribution is the unequal distribution of livestock. The lower the number of farmers concentrating on organic matter and its nutrient content from common grazing land on their compound fields, the higher the chance that the soil organic matter status can be maintained locally and the nutrient balance is positive. As stated above (in Section 8.3.2), comparing concentration of nutrients on marginal land with concentration of water in semiarid regions, redistribution of poverty creates opportunities that are absent at homogeneous distribution. This mechanism of reinforcement of redistribution is illustrated well by the study of Ramisch (2005) about "inequality, agro-pastoral exchanges, and soil fertility gradients"; the correlation between the N balance and stocking rate has been presented previously (in Section 8.3.2), and the disturbance by manure trade. Key assets for this trade and for focused application of manure, after passive concentration by livestock, appear to be carts and draft power. ... access to labor, transport, and land constrains efficient household manure use more than the number of livestock" (Ramisch, 2005, p. 366).

Ramisch asked farmers to classify themselves as "weak", "average", and "strong" regarding their soil fertility management ability, using their local criteria. These criteria appear to be accessibility of manure and inorganic fertilizers, availability of household labor and oxen, and timeliness of planting. The three classes show a heterogeneous distribution in the observed inverse relationship between the degree of heterogeneity of nutrient use and the household N input rate. The "weak" soil managers, having an average N input rate of 40 kg ha^{-1}, apply their inputs inequitably; the "strong" soil managers, having an average N input rate of 70 kg ha^{-1}, apply their inputs most equitably.

> Numerous comments made by farmers indicated that it is better to concentrate [scarce] resources than to scatter them. Households with scarce resources were therefore likely to concentrate it in a "hotspot" for maximum benefit, while households with access to more of that resource would either increase the number or the area of those "hotspots", thereby reducing the overall "patchiness" of their use. (Ramisch, 2005, p. 365)

One could state also that the land of "strong" soil managers becomes the fertile soil "hotspots" of the village territory.

8.3.5   The Limits of the Redistribution Phenomenon

Droughts, like those of the Sahel during the last three decades, are revelations regarding the degree of overpopulation at certain forms of land use and the surplus value of arable farming in comparison to livestock production for feeding men based on natural resources only. The drought in the overpopulated Sahelian countries changed production systems (Breman et al., 1990; de Grandi, 1996). The relative spatial separation of arable farming and livestock production disappeared, the mobility of pastoral systems decreased or even disappeared, the point of gravity of livestock production moved southward, from the Sahel to the savannah, and the integrated crop–livestock system became general. A change in the priority of goals for keeping livestock accompanied this change. The pastoral systems that dominated land use in the drier parts of Sahelian countries in the past aimed at producing milk and meat. In the present integrated crop–livestock systems, maintenance of crop production through manure, traction, and savings dominates over proper animal production. This difference becomes increasingly visible at an increasing stocking rate parallel to increased acreage under crops (see Section 8.2.3), which is caused by the decreasing average fodder quality. "Where the hogs are many, the wash is poor"; milk and meat production decrease more than proportionally with decreased fodder quality. At a certain moment, the average quality becomes too low to enable maintenance of the herd; the productivity parameters become too low. The fodder quality limits are 0.9% of N and 52% digestibility (Ketelaars, 1991).

For keeping as much livestock as possible and maintaining their productivity as well as possible, farmers are finally obliged to supplement their grazing animals. First, their own crop by-products are used, but finally part of their fields must be used for fodder production, or extra (high) quality fodder must be procured (Slicher van Bath, 1960; Mazoyer & Roudart, 1998). It will make a difference if former pastoralists, who still try to optimize animal production as a goal in itself, handle the crop–livestock system or if arable farmers who focus on crop production handle it. At first glance, the results of Ramisch (2005) seem counterintuitive in this context. Those who make the most effort to improve the fodder situation are arable farmers with rather limited livestock numbers; they collect cowpea leaves, cereal stover, and sweet potato vines. The original pastoralists do not do it. Their grazing strategy is, however, much more effective than those of arable farmers; the first group allows animals to graze longer and even night grazing may still be practiced (Leloup, 1994). The arable farmers must try increasing livestock numbers in spite of the "opposition" of existing well-managed herds of former pastoralists.

The best chances of rapidly increasing livestock density occur when there is an important market for the proper livestock products—milk and meat. The income obtained enables farmers to decrease the production of human food at the benefit of fodder production. This occurs, for example, in the neighborhood of cities (de Ridder & Slingerland, 2001). Livestock traders may also invest in improved feeding. Leloup (1994) presents the example of cotton seedcake procurement. Bieleman (1987) describes the rapid change in Drente in the nineteenth century of integrated systems dominated by rye production to those dominated by fodder production, triggered by the rapid increase of demand for milk and meat in industrializing Europe. This change facilitated the transition to inorganic fertilizer use, which in Drente appeared to be economically more feasible for livestock feeding than for rye production. In the Sahelian examples presented, inorganic fertilizers also start to play a role; primarily, cotton pays for it in this case.

The integrated crop–livestock system is indeed an effective tool for maintaining for a long time an increasing population based on natural resource use only. It has, however, its limits—a judgment also supported by those who accepted that the first analyses overestimated the negative trends in soil fertility decline (e.g., Turner, 1995; de Ridder et al., 2004; Ramisch, 2005). The limits are reached when the viability of livestock herds cannot be ensured any longer and/or the degradation of the depleted grazing land gains momentum. The chance that this happens increases rapidly when farmers start to reinforce the redistribution process provoked by their livestock, collecting plant biomass, mulch, or even topsoil from grazing land. In Drente, the collection of sod was a more important cause of desertification than livestock.

When the limits of integrated crop–livestock systems are reached depends on production systems and management decisions as described under Section 8.3.3. Those decisions vary with farmers' objectives and with their socioeconomic and policy environments. The number of factors involved and linkages between them are high. It is therefore difficult for farmers to make decisions and for others to advise them. Modeling and simulation is increasingly used to increase insight and to support the decision process; multiple goal planning is one of the instruments (e.g., Bakker et al., 1998; Sissoko, 1998; Savadogo, 2000; Stroosnijder & Van Rheenen, 2001; López-Ridaura, 2005). The use of inorganic fertilizers appears to be the tool to allow further population growth while reversing the soil depletion trend and allowing the unlinking of crop and livestock. Farmers' accessibility of input and output markets is key to the adoption of this solution; therefore, policymakers are as responsible for (un)sustainable land use as farmers.

8.4   Exploiting and Optimizing the Redistribution Phenomenon

8.4.1   Integrated Soil Fertility Management

One may wonder if inorganic fertilizer use can indeed be the tool for SSA to allow further population growth while reversing soil depletion. The market-oriented production in an increasingly globalizing world requires from farmers highly competitive production. The problem for regions with poor soils is, however, the unfavorable value-incremental yield–fertilizer cost ratio of inorganic fertilizer use (see Section 8.2.1). It is here that the redistribution phenomenon can be of use. The frequently heard and well-intentioned advice to use fertilizers on bush fields in view of the use of compost and manure on compound fields is like "putting the cart before the horse." The VCR of using inorganic fertilizer on compound fields, combining it with manure and/or compost, is higher than on bush fields with their negative organic matter and nutrient balances. The combined use of inorganic fertilizer with compost or manure is one of the effective forms of integrated soil fertility management (ISFM). ISFM, in this context, aims to improve access to, and increased use of, inorganic fertilizers. It combines inorganic fertilizers and soil amendments in an integrated way. Soil amendments—sources of organic matter, in particular, but sometimes also phosphate rock and/or lime, improve the soil organic matter status, the accessibility of P and the soil pH, and improve fertilizer-use efficiency. The inorganic fertilizer not only increases crop yield, it also contributes to the availability and quality of the key amendment, organic matter in form of crop by-products (IFDC, 2002, 2004, and 2005). Sissoko (1998) identifies a series of policy measures that favor the adoption of ISFM by cotton farmers in southeast Mali: those that increase the profitability of fertilizer use (see Section 8.3.1), soil improvement (e.g., supporting the use of soil amendments such as phosphate rock), improving land-use rights security, responsibility of producers through decentralization and development coordination. After 4 years of using ISFM by 3,000 farmers in seven West African countries on five crops, N-use efficiency increased by an average of 50% (IFDC, 2004). VCRs for traditional use varied between 2–3 or less[6] and 8, depending on systems, crops, and regions. The related yields varied between 750 kg ha^{-1} for maize on bush fields and 3,000 kg ha^{-1} for irrigated rice. After 4 years of using ISFM, VCR values ranged from 4 to 12, and yields from 1,800 to 5,500 kg ha^{-1}. The lowest VCR and yield was now found for sorghum, the highest again for irrigated rice. A whole menu of ISFM technologies has been developed; each technology has its own source of organic matter and its own recommendation domains. Details are increasingly published in scientific publications (Fofana et al., 2004, 2005, and in press; Wopereis et al., in press).

---

[6] These low values have been derived from on-farm trials; farmers generally do not adopt inorganic fertilizer use at this VCR level (see Section 8.3.1).

Good soils, enabling the economical use of inorganic fertilizers, are the most favorable for the transformation of extensive subsistence agriculture into intensive market-oriented production; they allow for competitive production. Relatively good soils exist everywhere in the form of compound fields.[7] IFDC compared fertilizer-use efficiency on compound and bush fields in the Sahel and the Soudanian savannah.

8.4.1.1 Millet Production on Compound and Bush Fields in Karabedji, Niger (Sahel; Average Rainfall 500 mm per year)

Without using fertilizer, in 3 successive years, millet grain yields on bush fields varied between 150 and 180 kg ha^{-1} and on compound fields between 490 and 570 kg ha^{-1}. Both N and P were limiting. Maximum N-use efficiencies were found by using doses of 30 kg ha^{-1} for both elements that led to average yields of 1,220 and 1,940 kg ha^{-1} for outlying and compound fields, respectively. Every kilogram of N produced 35 kg of millet grain on the bush fields compared with 47 kg on the compound field (Fofana et al., in press).

8.4.1.2 Maize Production on Compound and Bush Fields in Northern Togo (Soudanian Savannah; Average Rainfall 900 mm per year)

While comparable results have been obtained for individual sites and years, the average difference in N-use efficiency that has been observed on four farms in 3 successive years is less pronounced than for millet in the Sahel. Using doses of 50 kg ha^{-1} of N at 15 or 30 kg ha^{-1} of P produced maximum N efficiencies. Every kilogram of N produced 22 kg of millet grain on the bush fields compared with 25 kg on the compound field. The greatest differences between outfields and compound fields occurred during a year with low and erratic rainfall. For doses of 50 kg ha^{-1} of N, every kilogram of N produced 10 kg of maize grain on bush fields compared with 18 kg of maize grain on compound fields.

A much higher difference appears to exist between the N recovery, which is on average 30% higher on the compound fields in comparison with that of the bush fields. Different dose–N-uptake curves are found for the two types of fields. Yields increase proportionally with the N doses for the bush fields, but for the compound fields a saturation curve is found. In other words, N was not the limiting factor any more above 50 N, and absorbed N is increasingly less efficient when used for grain production (derived from Wopereis et al., in press).

---

[7] Market gardening around cities is in this context regarded as an extreme case of compound field production. It is the food transport for the urban population that replaces the centripetal concentration or organic matter by cattle, while organic urban waste replaces household waste and manure.

To better understand the differences in results for the Sahel and the savannah, one should realize that the C and N content of the savannah soils are about five times higher than those in the Sahel.[8] On the compound fields, in particular, important amounts of N became available from the soil; the uptake rate by crops was 50 kg ha^{-1} or more. Without any fertilizer, this leads to average maize yields of 2,100 kg ha^{-1} compared with 900 kg ha^{-1} on bush fields. The implication for the topic of this paper is that: (i) ISFM has indeed comparative advantages for marginal land (Breman, 1990); and (ii) that there is a limit to the benefits of redistribution. The latter appears also from the behavior of farmers in the villages studied by Ramisch (2005; see Section 8.3.3). It must be possible to improve the effects of N application in the savannah, but it requires detailed studies about limiting factors and, presumably, more sophisticated and more expensive fertilizer use, paying attention to more nutrients, time, and place of application, and fragmentation of doses.

The increased effects of soil improvement on fertilizer-use efficiency in the case of drought should be stressed more. Too often, it is suggested that the use of inorganic fertilizer is very risky in case of drought. The results obtained in Niger and the effects in a dry year in the savannah show that using ISFM can suppress this risk. This effect of integrated management of inorganic fertilizer and organic soil amendments appears far before the soil organic matter status is improved enough to explain the drought effect through a higher water-holding capacity (de Ridder & van Keulen, 1990). Other possible explanations are the organic matter contributions to improved water infiltration, the nutrient-holding capacity of the soil (see below), or the improved root development observed by Cissé (1986).

8.4.2 Optimizing Redistribution

Use and optimum management of (soil) organic matter must be different in case of being the main source of nutrients or being the soil amendment for efficient inorganic fertilizer use (e.g., Palm et al., 2001). Although the interactions between inorganic fertilizers and organic matter are not yet known in detail, several processes are known through which organic matter contributes to effective management of inorganic macronutrient fertilizers (Vanlauwe et al., 2002). Besides the above-mentioned improvement of water absorption potential, it concerns processes that lead to the improvement of the nutrient absorption potential: increased cation exchange capacity (CEC) and increased anion exchange capacity, pH buffering, occupation of phosphate fixation sites, and maintenance of micronutrient balances through chelation and ion exchange.

---

[8] In the Sahelian case the top soils of compound and bush fields had an organic C content of 1.6 and 1.5 g kg^{-1}, respectively, and a total N content of 135 and 118 mg kg^{-1}. For the savanna case these figures were 13.4 and 6.3 for C and 968 and 511 for N.

The mechanisms behind these phenomena are such that it appears sufficient to ensure that minimum soil organic matter values are maintained through organic matter management. Pieri (1989) formulates such critical threshold values in relation to the soil textures, paying attention to the soils' potential for erosion. Breman (2002), influenced by the work of Janssen et al. (1990), suggests a soil organic matter content in relation to the nutrient-holding capacity of soils; the CEC value should be at least 10 cmol kg^{-1}. Also, in that case, the threshold and the required amounts of organic matter will be texture dependent, considering the contribution of clay particles to the CEC. The idea of a threshold is supported by the behavior of farmers as described by Ramisch (2005, p. 365), presented in Section 8.3.3. Those households that "would either increase the number or the area of those "hotspots," thereby reducing the overall "patchiness" of their use" are the richer ones using also most inorganic fertilizers.

Not only is less organic matter required when used as a soil amendment in an ISFM context compared with its use as manure, but its quality is also different. Organic matter with a lower nutrient content and a lower mineralization rate is required (Palm et al., 2001). This concerns material that contributes more easily and quickly to soil organic matter maintenance or formation; it has a high C-sequestration capacity. The availability of such organic matter is much higher than that of the quality class that can be used as a direct alternative for inorganic fertilizer. However, part of the organic matter will be "too inert" to serve effectively in an ISFM context in combination with inorganic fertilizers (Henkens, 1975; Palm et al., 2001; Breman et al., 2004). This organic matter can only be used for erosion control and for C-sequestration.

Armed with this knowledge and with the knowledge of farmers and scientists about spatial crop growth variability (e.g., Voortman & Brouwer, 2003; Voortman et al., 2004), the progressive transition from passive to active redistribution can be guided and exploited for the intensification of agriculture by inorganic fertilizer use. Goal- and location-specific recommendations can be made for the integrated use of inorganic fertilizers and available sources of organic matter, promoting different ISFM technologies for different farm sites[9]:

- One should be less concerned about negative organic matter and nutrient balances of whole village territories and adjacent rangelands, and even about the degradation of the latter. Rather, one should be concerned about bottlenecks for market-oriented, competitive production, such as limited access to inorganic fertilizers, lack of market transparency, and inadequate agricultural policies.
- Intensification should start on compound fields[5] and focus first on N fertilizer, combining it with manure and compost, and if required with P (K, S, ...).
- Compound fields on which inorganic fertilizer is used can be extended more than the traditional ones on which manure and compost serve as nutrient sources.

[9] It goes without saying that besides benefiting from the (relative) good soils of compound fields for starting intensification, farmers should also exploit good soils such as those of depressions and valleys, where nutrients from elsewhere are concentrated thanks to water. Schreurs et al. (2002) promote and illustrate the use of strategic sites in general.

- At least part of the space[10] created by extension of compound fields should be used for fodder production on bush fields. If the VCR of fertilizer use allows it, other ISFM technologies can be introduced on bush fields, particularly those based on agroforestry, fodder, and cover crops, and pasture–crop rotation. It reinforces the redistribution process and the availability of organic matter for the extension of compound fields without (rapid) depletion of bush fields. Both N and P are required (more often than on compound fields—K, S, and micronutrients, e.g., for productive leguminous crops).
- N doses must be relatively low to avoid the negative interaction with N from soil organic matter and other organic resources that easily suppress the benefit of high N recovery (Fofana et al., in press; Wopereis et al., in press).

Small ruminants, the most important tool for organic matter *casu quo* mineral nutrient transport where population pressure is high and depletion of forest-, range-, and wasteland is advanced, become redundant, and their harmful influence on the environment can be controlled when the redistribution phenomenon itself is not used to maintain production but where it serves intensification as described. The centripetal transport can even turn into a centrifugal transport and replenish (planned or unplanned) the depleted surroundings of farms and villages. A form of unplanned redistribution in countries with intensive agriculture and high levels of fertilizer use is acid rain, a fraction of which is caused by extremely intensive livestock raising. In the Netherlands, its effect on entirely depleted and degraded land reached the point that the law had to be changed to protect the last moving dunes as a monument to the history of agriculture. These laws, very similar to those developed today for stopping desertification in the Sahel, were elaborated centuries ago to stop environmental degradation caused by human activities in regions with poor sandy soils (e.g., Gelderland Province, 1862; see Handelingen Provinciale Staten Gelderland, 1982).

8.4.3   Enabling Socioeconomic and Policy Environments

Active replenishment of depleted African soils is no requirement for agricultural development. However, public investments in soils can contribute largely to the success of ISFM. Such investments are about one-tenth of those in small-scale irrigation with at least comparable rates of return and can be direct (cheap soil amendments) or indirect—supporting farmers through subsidies or loans for carts and animal traction (Breman et al., 2003). They should focus on reinforcement of the positive effects of the redistribution phenomenon, enabling farmers to start fertilizer use where even the compound fields at present do not allow it.[11] It forms

---

[10] 'Space' should be read in several senses: more food can be produced thanks to higher yields and larger surface, while the higher fertilizer-use efficiency leads to a higher net production and income.

[11] A useful form of support concerns reinforcement of active organic matter management and redistribution, for example, through subsidized carts and draught oxen.

one of the four main public supports for agricultural and rural development. The others are investments in infrastructure and transport, the creation of an enabling environment for private input and output market development (transparency), and regional cooperation, considering the benefits of scale and temporary and differentiated market protection (Breman & Debrah, 2003). Such policies improve the VCR of using inorganic fertilizer by offering farmers lower fertilizer prices and higher crop prices, which reinforce the approach presented for increased fertilizer-use efficiency.

Acknowledgments The authors gratefully acknowledge Tom Crawford, Bert Meertens, Florus van der Pol, Pieter Pypers, and Coen Reijntjes for their comments and suggestions and Marie Thompson for editing.

References

Bakker, E. J., Quak, W., Sissoko, K., Moh, S. M. T., & Hengsdijk H. (1998). Vers une agriculture durable. In H. Breman & K. Sissoko (Eds.), *L'intensification agricole au Sahel* (pp. 125–141). Paris: KARTHALA.

Bieleman, J. (1987). *Boeren op het Drentse zand, 1600–1910. Een nieuwe visie op de oude landbouw.* Utrecht, The Netherlands: H&S Uitgevers.

Binswanger, H., & Pingali, P. (1988). Technological priorities for farming in sub-Saharan Africa. *World Bank Research Observations, 3,* 81–98.

Breman, H. (1990). No sustainability without external inputs. In *Beyond adjustment.* Africa seminar, Maastricht, the Netherlands (pp. 124–133). Den Haag: Ministry of Foreign Affairs.

Breman, H. (1998). Soil fertility improvement in Africa, a tool for or a by-product of sustainable production? Special Issue on Soil Fertility. *African Fertilizer Market, 11*(5), 2–10.

Breman, H. (2000). Sustainable agricultural intensification in Africa: The role of capital. Lecture African Development Bank, June 6th, Abidjan. Lomé: Internal Report IFDC-Africa.

Breman, H. (2002). Soil fertility and farmers' revenue: Keys to desertification control. In H. Shimizu (Ed.), *Integration and regional researches to combat desertification; present state and future prospect* (pp. 26–41). Tsukuba, Japan: The 16th Global Environment Tsukuba. CGER/NIES.

Breman, H. & Sissoko, K. (Eds.) (1998). *L'intensification agricole au Sahel.* Paris: KARTHALA.

Breman, H., & Reuler, H. van (2002). Legumes, when and where an option? No panacea for poor tropical West African soils and expensive fertilizers. In B. Vanlauwe, J. Diels, N. Sanginga, & R. Merckx (Eds.), *Integrated plant nutrient management in sub-Saharan Africa* (pp. 285–298). Wallingford, UK: CABI.

Breman, H., & Debrah, S. K. (2003). Improving African food security. *SAIS, 23*(1), 153–170.

Breman, H., Ketelaars, J. J. M. H., & Traoré, N. (1990). *Un remède contre le manque de terre? Bilan des éléments nutritifs, production primaire et élevage au Sahel. Sécheresse, 2,* 109–117.

Breman, H., Gakou, A., Mando, A., & Wopereis, M. C. S. (2004). Enhancing integrated soil fertility management through the Carbon Market to combat resource degradation in overpopulated Sahelian countries. Proceedings of Regional Scientific Workshop on Land Management for Carbon Sequestration. Bamako (Mali), February 27–28 (2004). Organized by the Carbon from Communities project and funded by the U.S. National Aeronautics and Space Administration. IER (Bamako), NASA and Soil Management CRSP, USA. CD-Rom NASA.

Breman, H., Wopereis, M. C. S., Maatman, A., Hellums, D., Chien, S. H., & Bowen, W. (2003). Food security and economic development in sub-Saharan Africa: Investments in soils or in irrigation? Presentation AEPS workshop, USAID, September 22–26, Rosebank (South Africa). Internal report IFDC, Lomé.

Buresh, R. J., Sanchez, P. A., & Calhoun, F. (Eds.) (1997). *Replenishing soil fertility in Africa.* SSSA Special Publication No. 51, SSSA, Nairobi: Medison and ICRAF.

Cissé, L. (1986). Etude des effets d'apports de matière organique sur les bilans hydriques et minéraux et la production du mil et de l'arachide sur un sol sableux dégradé du Centre-Nord du Sénégal. Thèse de Doctorat.

Cour, J. M. (2001). The Sahel in West Africa: Countries in transition to a full market economy. *Global Environmental Change, 11*, 31–47.

Duivenbooden, van, N. (1992). Sustainability in terms of nutrient elements with special reference to West-Africa. Cabo-Dlo, Report Wageningen, The Netherlands 160. 261 pp.

Fofana, B., Breman, H., Carsky, R. J., Reuler, H. van, Tamélokpo, A. F., & Gnakpenou, K. D. (2004). Using mucuna and P fertilizer to increase maize grain yield and N fertilizer use efficiency in the coastal savannah of Togo. *Nutrient Cycling in Agroecosystems, 68*, 213–222.

Fofana, B., Tamélokpo, A. F., Wopereis, M. C. S., Breman, H., Dzotsi, K., & Carsky, R. J. (2004). Nitrogen use efficiency by maize as affected by a mucuna short fallow and P application in the coastal savanna of West Africa. *Nutrient Cycling in Agroecosystems, 69*, 1–10.

Fofana, B., Wopereis, M. C. S., Bationo, A., Breman, H., Tamélokpo, A. F., Gnakpénou, K. D., Ezui, K., Zida, Z., & Mando, A. (in press). Millet nutrient use efficiency as affected by inherent soil fertility in the West African Sahel.

Garin, P., Faye, A., Lericollais, A., & Sissokho, M. (1990). Evolution du rôle du bétail dans la gestion de la fertilité des terroirs Sereer au Sénégal. *Les Cahiers de Recherche Développement, 26*, 65–94.

Grandi, J. C. de (1996). *L'évolution des systèmes de production agropastorale par rapport au développement rural durable dans les pays d'Afrique soudano–sahélienne.* Collection FAO: Gestion des exploitations agricoles no. 11. Rome: FAO.

Greene, L. W., & Cole, N. C. (2006). Feedlot nutrient management impacts efficiency and environmental quality. Available at http://www.admani.com/alliancebeef/TechnicalEdge/FeedlotNutrientManagement.htm

Haileslassie, A. (2005). *Soil nutrient balance at different spatial scales: examining soil fertility management and sustainability of mixed farming systems in Ethiopia.* Dissertation. Göttingen, Germany: University of Goettingen, Cuvillier Verlag.

Henao, J. & Baanante, C. (1999). Estimation rates of nutrient depletion in soils of agricultural lands of Africa. Technical Bulletin IFDC-T-48. Muscle Shoals: IFDC.

Henao, J. & C. Baanante, C. (in press). *Agricultural production and soil nutrient mining in Africa: Implications for resource conservation and policy development.* Paper-R7. Muscle Shoals, AL: IFDC.

Henkens, Ch. H. (1975). Brengt de gangbare landbouw schade toe aan bodemvruchtbaarheid? *Bedrijfsontwikkeling, 6*(3), 207–214.

Hilhorst, T., & Muchena, F. (Eds.) (2000). *Nutrients on the move. Soil fertility dynamics in African farming systems.* London: IIED.

IFDC. (2002). Collaborative research programme for soil fertility restoration and management in resource-poor areas of sub-Saharan Africa. IFDC-Africa (Lomé); TSBF (Nairobi) and ACFD (Harare), supported by IFAD (Rome). Technical Bulletin IFDC-T-67. Muscle Shoals, AL: IFDC.

IFDC. (2004). The integrated soil fertility management project. IFDC-Africa (Togo), DGIS, Ministry of Foreign Affairs (The Netherlands), IFAD (Rome), IFA (Paris) and USAID (Washington). Muscle Shoals, AL: IFDC.

IFDC. (2005). Development and dissemination of sustainable integrated soil fertility management practices for small holder farms in sub-Saharan Africa. IFAD Technical Assistance Grant no. R535, Final Report. TSBF Institute of CIAT, Nairobi and IFDC-Africa, Lomé.

Janssen, B. H., Guiking, F. C. T., van der Eijk, D., Smaling, E. M. A., Wolf, J. & van Reuler, H. (1990). A system for quantitative evaluation of the fertility of tropical soils (QUEFTS). *Geoderma, 46*, 299–318.

Kelly, V., Jayne, T., & Crawford, E. (2005). Farmers' demand for fertilizer in Sub-Saharan Africa. Draft paper prepared for the World Bank by the Dept. of Agric. Economics, East Lansing: Michigan State University. Available at www.nrinternational.co.uk/uploads/documents/BP2VALPg360205.pdf

Ketelaars, J. J. M. H. (1991). La production animale. In H. Breman & N. de Ridder (Eds.), *Manuel sur les pâturages des pays sahéliens* (pp. 357–389). Paris: ACCT, Paris: CTA, Wageningen, The Netherlands: KARTHALA.

Keulen, H. van, & Breman, H. (1990). Agricultural development in the West African Sahelian region: A cure against land hunger? *Agriculture, Ecosystems & Environment, 32*, 177–197.

Krul, J. M., Penning de Vries, F. W. T., Stroosnijder, L., & Pol, F. van der (1982). Le phosphor dans le sol et son accessibilité aux plantes. In F. W. T. Penning de Vries & M. A. Djitèye (Eds.), *La productivité des pâturages sahéliens. Une étude des sols, des végétations et de l'exploitation de cette ressource naturelle* (pp. 246–283). Agricultural Research Report 918, PUDOC, Wageningen, The Netherlands.

Lekasi, J. K., & Kimani, S. K. (2005). Livestock management and manure quality. In C. E. N. Saval, M. N. Omare & P. L. Woomer (Eds.), *Organic resource management in Kenya. Perspectives and guidelines*, Chapter 2. Available at www.formatkenya.org.

Leloup, S. (1994). *Multiple use of rangelands within agropastoral systems in southern Mali*. Ph. D. thesis, Wageningen, The Netherlands: Agricultural University Wageningen.

López-Ridaura, S. (2005). *Multi-scale sustainability evaluation. A framework for the derivation and quantification of indicators for natural resource management systems*. Ph.D. Thesis, Wageningen, The Netherlands: Wageningen University.

Mazoyer, M., & Roudart, L. (1998). Histoire des agricultures du monde du Néolithique à la crise contemporaine. du SEUIL (Eds.), Paris.

Mazzucato, V., & Niemeijer, D. (2000). *Rethinking soil and water conservation in a changing society: A case study in eastern Burkina Faso*. Tropical Resource Management Paper 32, Wageningen, The Netherlands: Wageningen University and Research Center.

Meertens, B. (2005). A realistic view on increasing fertilizer use in sub-Saharan Africa. Available at www.meertensconsult@hetnet.nl

MINECOFIN (2003). *Rwanda development indicators*. Statistics Department, Ministry of Finance and Economic Planning. Kigali, Rwanda: Pallotti Presse.

Palm, C. A., Gachengo, C. N., Delve, R. J., Cardish, R. J., & Giller, K. E. (2001). Organic inputs for soil fertility management in tropical agroecosystems. Application of an organic resource database. *Agriculture, Ecosystems & Environment, Special Issue 71*, 255–267.

Penning de Vries, F. W. T., & Djitèye, M. A. (Eds.) (1982). *La productivité des pâturages sahéliens. Une étude des sols, des végétations et de l'exploitation de cette ressource naturelle*. Agricultural Research Report 918. Wageningen, The Netherlands: PUDOC.

Pieri, C. (1989). Fertilité des terres de savanes. Bilan de trente ans de recherche et de développement agricol au Sud du Sahara. Montpellier, France: Min.de la Coopération et CIRAD-IRAT.

Pol, F. van der (1992). *Soil mining: An unseen contributor to farm income in Southern Mali*. Bulletin 325, Amsterdam: Royal Tropical Institute (KIT).

Quak, W., Hengdijk, H., Touré, M. S. H., Sissoko, K., Camara, O., Dembélé, N. F., & Bakker, E. J. (1998). Activités de production agricole durable. In H. Breman & K. Sissoko (Eds.) (1998). *Intensification agricole au Sahel* (pp. 539–561). Paris: KARTHALA..

Ramisch, J. J. (2005). Inequality, agro-pastoral exchanges, and soil fertility gradients in southern Mali. *Agriculture, Ecosystems & Environment, 105*, 353–372.

Ridder, N. de, Breman, H., Keulen, H. van & Stomph, T. J. (2004). Revisiting a 'cure against landhunger': Soil fertility management and farming systems dynamics in the West African Sahel. *Agricultural Systems, 80*, 109–131.

Ridder, N. De., & Keulen, H. van (1990). Some aspects of the role of organic matter in sustainable intensified arable farming systems in the West African semi-arid tropics (SAT). *Fertilizer Research, 26*, 299–310.

Ridder, N. de, & Slingerland, M. (2001). Reflections on interventions by land users in Sahelian villages. In L. Stroosnijder & T. van Rheenen (Eds.), Agro-silvo-pastoral land use in Sahelian villages. *Advances in Geoecology, 33*, 297–302.

Smaling, E., Toure, M., Ridder, M, N. de, Sanginga, N., & Breman, H. (in press). Fertilizer use and the environment in Africa: Friends or foes? Background paper presented for the African Fertilizer Summit 9–13th June (2006), Abuja, Nigeria. NEPAD, Johannesburg/IFDC, Muscle Shoals. Proceedings Africa Fertilizer Summit.

Turner, M. (1995). The sustainability of crop-livestock systems in sub-Sahara Africa: Ecological and social dimensions neglected in the debate. In J. M. Powell, S. Fernández-Rivera, T. O. Williams & C. Renard (Eds.), *Livestock and sustainable nutrient cycling in mixed farming systems of sub-Saharan Africa* (pp. 435–452). *Vol. II.* Technical papers. Proceedings of an international conference. 22–26 November 1993. Addis Ababa: ILCA.

Samaké O. (2003). *Integrated crop management strategies in Sahelian land use systems to improve agricultural productivity and sustainability: A case study in Mali.* Tropical Resource Management Papers 47. Wageningen, The Netherlands: Wageningen University and Research Center.

Savadogo, M. (2000). *Crop residue management in relation to sustainable land use. A case study in Burkina Faso.* Tropical Resource Management Papers 31. Wageningen, The Netherlands: Wageningen University and Research Center.

Schreurs, M. E. A., Maatman, A. & Dangbégnon, C. (2002). In for a penny in for a pound. Strategic site-selection and on-farm client-oriented research to trigger sustainable agricultural intensification. In B. Vanlauwe et al. (Eds.), *Integrated plant nutrient management in sub-Saharan Africa* (pp. 63–94). Wallingford, UK: CABI.

Sissoko, K. (1998). Et demain l'agriculture? Options techniques et mesures politiques pour un développement agricole durable en Afrique subsaharienne. Cas du Cercle de Koutiala en zone sud du Mali. Tropical resource management papers 23. Wageningen, The Netherlands: Université Agronomique Wageningen.

Slicher van Bath, B. (1960). De agrarische geschiedenis van West-Europa 500–1850. Aula 565. Utrecht, Germany: Het Spectrum.

Stoorvogel, J. J., & Smaling, E. M. A. (1990). *Assessment of soil nutrient depletion in sub-Saharan Africa, 1983–2000. Report 28, Vol. 1.* Wageningen, The Netherlands: Winand Staring Centre (SC-DLO).

Stroosnijder, L., & Rheenen, T. van (Eds.) (2001). Agro-silvo-pastoral land use in Sahelian villages. *Advances in Geoecology, 33.*

Tiffen, M., Mortimore, M., & Guchuki, F. (1994). *More people, less erosion. Environmental recovery in Kenya.* Chichester, UK: Wiley.

Vanlauwe, B., Diels, J., Sanginga, N., & Merckx, R. (Eds.) (2002). Integrated plant nutrient management in sub-Saharan Africa. Wallingford, UK: CABI.

Voortman, R. L., & Brouwer, J. (2003). An empirical analysis of the simultaneous effects of nitrogen, phosphorus and potassium in millet production on spatially variable fields in SW Niger. *Nutrient Cycling in Agroecosystems, 66,* 143–164.

Voortman, R. L., Brouwer, J., & Albersen, P. J. (2004). Characterization of spatial soil variability and its effect on millet yield on Sudano-Sahelian coversands in SW Niger. *Geoderma, 121,* 65–82.

Wiggins, S. (1995). Change in African farming systems between the mid-1970s and the mid-1980s. *Journal of International Development, 7,* 807–848.

Wit, C. T. de (1986). Introduction. In H. van Keulen & J. Wolf (Eds.), Modelling of agricultural production : Weather, soils and crops (pp. 3–10). Simulation monographs. Wageningen, The Netherlands: PUDOC..

Wopereis, M. C. S., Tamélokpo, A., Ezui, K., Gnakpénou, D., Fofana, B., & Breman, H. (in press). Mineral fertilizer management of maize on farmer fields differing in organic inputs in the West African savanna. Field Crops Research.

Chapter 9
Soil Quality and Methods for its Assessment

Diego De la Rosa and Ramon Sobral

Abstract Environmental sustainability will only be achieved by maintenance and improvement of soil quality. Soil quality is considered as the capacity of a soil to function. Its assessment focuses on dynamic aspects to evaluate the sustainability of soil management practices. In this chapter, a wide perspective of soil quality and the complex task of its assessment, considering the inherent and dynamic factors, are introduced. It focuses on the possibilities of applying and integrating the accumulated knowledge in agroecological land evaluation in order to predict soil quality. Advanced information technologies in modern decision support tools enable the integration of large and complex databases, models, tools, and techniques, and are proposed to improve the decision-making process in soil quality management. Although universal recommendations on soil quality and sustainability of soil management must not be done, this chapter presents general trends in soil quality management strategies. This includes arable land identification, crop diversification, organic matter restoration, tillage intensity, and soil input rationalization.

Keywords Agroecological land evaluation, dynamic soil quality, inherent soil quality, MicroLEIS, soil function, soil health, soil indicator, spatial decision support tool, sustainable agricultural system

9.1 Introduction

The results of exploiting land-use systems without consideration of the consequences on soil quality have been environmental degradation. Agricultural use and management systems have been generally adopted without recognizing consequences on soil conservation and environmental quality, and therefore significant decline in agricultural soil quality has occurred worldwide (e.g., Imeson et al., 2006). Soil erosion and diffuse soil contamination are the major degradation processes on agricultural lands as a consequence of expansion and intensification of agriculture. Other nonagricultural uses, such as industrial and urban uses, also have

A.K. Braimoh and P.L.G. Vlek (eds.), *Land Use and Soil Resources.*
© Springer Science+Business Media B.V. 2008

important negative consequences on soil quality, due to local contamination, soil sealing, and changes in the dynamics of the landscape systems (see Chapter 10).

The concept of soil quality (Doran & Jones, 1996; Karlen et al., 1997) is useful to assess the condition and sustainability of soil and to guide soil research, planning, and conservation policy. However, some authors (e.g., Sojka & Upchurch, 1999; who first introduced this controversy) consider the soil-quality paradigm as parochial, despite its application by several US institutions such as the Soil Quality Institute. Other workers such as Davidson (2000) noted that soil quality is a valid and important concept that is not amenable to a simple and universal definition, and that will make a distinctive and crucial contribution to soil management.

The importance of soil quality lies in achieving sustainable land use and management systems, to balance productivity and environmental protection. Unlike water and air quality, simple standards for individual soil-quality indicators do not appear to be sufficient because numerous interactions and trade-offs must be considered. For assessing soil quality a complex integration of static and dynamic chemical, physical, and biological factors need to be defined in order to identify different management and environmental scenarios. Also, the consequences of any decline in soil quality may not be immediately experienced. The soil system does not necessarily change as a result of changing external conditions or use, because soil has the capacity of resistance (or resilience) to the effects of potentially damaging conditions or misuse or to filter out harmful materials added to it. In part, this capacity of the soil in buffering the consequences of inputs and changes in external conditions arises because the soil is an exceedingly complex and varied material with many diverse properties and interactions between soil properties. It is this complex dynamic nature which often makes it difficult to distinguish between changes as a result of natural development and changes due to nonnatural external influences. Soil-quality assessment, based on inherent soil factors and focusing on dynamic aspects of soil system, is an effective method for evaluating the environmental sustainability of land use and management activities (Nortcliff, 2002).

However, the process of evaluating soil is not new, and agroecological land evaluation has much to offer. Land suitability is defined in land evaluation as "the fitness of a given land unit for a specified type of land use" (FAO, 1976). In a more operational sense, suitability expresses how well the biophysical potentialities and limitations of the land unit match the requirements of the land-use type. Therefore, new investigations must obviously be based on a solid understanding of past studies (De la Rosa, 2005). Agroecological land evaluation predicts land behavior for each particular use, and soil-quality evaluation predicts the natural ability of each soil to function. However, land evaluation is not the same as soil-quality assessment, because biological parameters of the soil are not considered in land evaluation. Soil surveys are the building blocks of the dataset needed to drive land evaluation. Soil surveys and soil taxonomy systems are used to define with precision specific soil types.

Emerging technologies in data and knowledge engineering are providing excellent possibilities for the development and application processes of soil-quality assessment. As in land evaluation, the application phase of soil-quality assessment is a complex process of scaling-up from the representative areas of the development phase to implementation in unknown scenarios. This application phase can be executed with computer-assisted procedures. It involves the development and linkage of integrated components, the recently named decision or planning support tools (e.g., MicroLEIS system; De la Rosa et al., 2004). Decision support systems are computerized technology that can be used to support complex decision-making and problem-solving. Technically, decision support system comprises components for (i) sophisticated database management capabilities with access to internal and external data, information, and knowledge, (ii) powerful modeling functions accessed by a model management system, and (iii) simple user interface designs that enable interactive queries, reporting, and graphing functions (e.g., Oxley et al., 2004).

As reported by the Soil Quality Institute (USDA, 2006), the ultimate purpose of assessing soil quality is not to achieve high aggregate stability, biological activity, or some other soil property. The purpose is to protect and improve long-term agricultural productivity, water quality, and habitats of all organisms including people. By assessing soil quality, a land manager will be able to determine if a set of management practices is sustainable. For example, agricultural management systems located on the most suitable lands, according to their agroecological potentialities and limitations, are the best way to achieve sustainability.

There is a need to investigate coordinated and multidisciplinary approaches to assessing soil quality, evaluating long-term potential and limitations (inherent soil aspects), and monitoring the short-term changes (dynamic soil aspects) in response to sustainable soil use and management. This chapter presents a wide perspective on soil quality and the complex task of its assessment, considering the inherent and dynamic aspects of soil system. It focuses on the possibilities for applying and integrating accumulated knowledge on land-evaluation modeling, in order to predict soil-quality indexes. Advanced information technologies, which enable the integration of large and complex databases, models, tools and techniques, are proposed to improve the decision-making process in soil-quality assessment application. Finally, general trends in soil-quality management strategies are discussed.

9.2 Soil Quality

As suggested in the early 1990s, soil quality is "the capacity of a soil to function". More specifically, soil quality has been defined by a committee for the Soil Science Society of America (Karlen et al., 1997) as "the capacity of a specific kind of soil to function, within natural or managed ecosystem boundaries, to sustain plant and animal productivity, maintain or enhance water and air quality, and support human health and habitation". Also, soil quality can be considered as the ability of a soil to fulfill its functions in the ecosystem, which are determined by the integrated actions of different soil properties. With respect to agriculture, soil quality would be the soil's fitness to support crop growth without becoming degraded or otherwise harming the environment.

Some authors (e.g. Warkentin, 1995) have suggested that soil quality is simply related to the quantity of crops produced. However, others have emphasized the importance of demonstrating how soil quality affects feed and food quality, or how soil quality affects the habitat provided for a wide array of biota. Numerous other aspects associated with the living and dynamic nature of soil will be encountered if the concept of soil quality is considered in relation to different land uses: forest and rangeland ecosystems, urban and industrial land, recreational uses, etc. Because of the diversity of potential land uses, the concept of soil quality should be viewed as relative rather than absolute. Therefore, each soil has a natural capacity to perform a specific function.

According to the Soil Quality Institute (USDA, 2006), the soil-quality concept is related to the concepts of sustainability of soil use and management, although in some cases the focus has been predominantly on contaminated land. To do that the notion of soil quality must include soil productivity, soil fertility, soil degradation, and environmental quality. In this sense, the major activity is devoted to the evaluation of sustainable soil management systems together with the development of associated soil-quality assessments (Doran & Jones, 1996).

9.2.1   Quality Types

Soil has both inherent and dynamic qualities (USDA, 2006). Inherent soil quality is a soil's natural ability to function. For example, sandy soil drains faster than a clayey one. Deep soil has more room for roots than soils with bedrock near the surface. These characteristics are permanent and do not change easily. The inherent quality of soils is often used to compare the abilities of one soil against another, and to evaluate the value or suitability of soils for specific uses. Traditional studies in land evaluation have been basically concerned with the practical interpretation of inherent soil properties (soil suitability) such as inventoried in soil surveys.

Dynamic soil quality is how soil changes depending on how it is managed. Management choices affect the amount of soil organic matter, soil structure, and water- and nutrient-holding capacity. One goal of soil-quality research is to learn how to manage soil in a way that improves its functions. This dynamic aspect of soil quality is the focal point of assessing and maintaining healthy soil resources.

According to the soil factors considered, the soil quality can be physical, chemical, or biological. Most of the physicochemical factors are related to inherent soil quality, and biological and some physical factors with the dynamic soil quality. Although soil quality often focuses on biological aspects, this must not diminish the importance of physical and chemical factors (Ball & De la Rosa, 2006).

9.2.2   Soil Health

Soil health is the other principle for sustainable soil management used by some soil scientists. Doran et al. (1997) define soil health as the continued capacity of soil to function as a vital living system, within ecosystem and land-use boundaries; to sustain biological productivity; promote the quality of air and water environments; and maintain plant, animal, and human health. In this sense, the soil is considered as a living system, address all essential functions of soil in the landscape, compare the condition of a given soil against its own unique potential within climatic, landscape, and vegetation patterns, and somehow enable meaningful assessments to trends. Although some authors consider the terms soil quality and soil health as synonymous (e.g., Wolfe, 2006), the integrated concept of soil quality can be defined, including the inherent soil quality, traditionally named soil suitability, and the dynamic soil quality or soil health (Fig. 9.1).

9.2.3   Soil Functions

The soil system can perform many functions, and often simultaneously. According to Nortcliff (2002), the soil must provide the following basic functions: (i) a physical, chemical, and biophysical setting for living organisms; (ii) the regulation and partition of water flow, storage, and recycling of nutrients and other elements; (iii) support for biological activity and diversity for plant growth and animal productivity; (iv) the capacity to filter, buffer, degrade, immobilize, and detoxify organic and inorganic substances; and (v) provide mechanical support for living organisms and their structures.

Specific soil functions can be defined with respect to issues like particular crop growth, and soil erosion or soil contamination hazard (Table 9.1).

Several soil physical functions, such as water retention and infiltration or soil aeration, are directly connected to the biological status of soil system, as also are the kinds of organisms and nutrient supply. Soil quality is therefore a multifunctional concept. As reported by Imeson et al. (2006), it is well known that overuse or exploitation of some functions (e.g., production function for crops) can lead to the damage of other ones. The spatial and temporal variation in the provision of functions should be incorporated in evaluations or assessments.

Soil quality

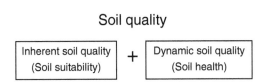

Fig. 9.1 Graphical representation of the soil-quality concept integrating inherent soil quality (or soil suitability) and dynamic soil quality (or soil health)

Table 9.1 Specific soil functions considered for several soil-quality issues

Soil-quality issue	Soil function
Crop growth	Plant root penetration
	Plant water-use efficiency
	Water- and air-filled pore space
	Water infiltration
Natural fertility	Nutrient availability
	Cation-exchange capacity
	Acidity
	Salinity/alkalinity
	Toxicity
Erosion risk	Runoff potential
	Erodibility
	Cover protection
	Subsoil compaction
	Workability
Compaction risk	Water retention
	Water infiltration
	Cohesion
	Workability/trafficability
Contamination risk	Leaching potential
	Toxic absorption
	Toxic mobility
	Chemicals degradation

9.2.4 Soil Threats

Consideration of soil threats is crucial for assessing the quality of the soil system. These are the major threats faced by soils: (i) soil erosion, (ii) soil contamination, (iii) decline in organic matter and biodiversity, (iv) soil compaction, (v) salinization, (vi) floods and landslides, and (vii) soil sealing. In many places, soil erosion is the most severe consequence of soil degradation with respect to restoration of soil quality, and controlling erosion is a prerequisite for a healthy soil. However, most of the soil degradation processes are interlinked, and are often linked by similar causative factors. The risk of these soil threats can be monitored by use of indicators such as trends in yields on soils under irrigation to monitor risk of salinity. Actions to protect soil quality necessitate tackling collectively the different threats.

Imeson et al. (2006) provide interesting information of the SCAPE (Soil Conservation and Protection for Europe) project, at different levels of scale, about soil degradation processes and how they are related to soil use and management. Ten case studies are reported by these authors which deal with the main threats to soils in different biogeographic regions in Europe. The general consensus of these case studies is that soil conservation and protection requires a holistic interdisciplinary approach and that integrated actions are required considering all of soil functions. The main task of the SCAPE project was to provide scientific support to the development of the European Soil Strategy to manage soils in a sustainable way and protect them from the many threats they are facing (EC, 2002).

9.3 Assessment Procedures

Any evaluations of soil quality must consider the multiple soil uses (e.g., agricultural production, forest, rangeland, nature conservation, recreation, or urban development). However, the most widely accepted concept of soil quality and the most significant in a global context concerns agro-ecosystems. In soil-quality evaluation or assessment, the two main questions that must be answered are: (i) how does the soil function; and (ii) what procedures are appropriate for making the evaluation. After answering those questions, a range of parameter values or indexes that indicate a soil is functioning at full potential can be calculated using landscape characteristics, knowledge of pedogenesis, and a more complete understanding of the dynamic processes occurring within a soil. Soil-quality assessment focuses on dynamic aspects to evaluate the sustainability of soil management practices, but it must be based on the inherent soil factors.

9.3.1 Soil-Quality Indicators

A soil-quality indicator is a simple attribute of the soil which may be measured to assess quality with respect to a given function. It is important to be able to select attributes that are appropriate for the task, given the complex nature of the soil and the exceptionally large number of soil parameters that may be determined, as exemplified in Table 9.2.

Table 9.2 Original geo-referenced soil-profile attributes (morphological, physical, and chemical properties) stored in the SDBmPlus soil database. (Adapted from De la Rosa et al., 2004)

Data block	Stored soil profile variables	
	Land characteristic type	Number[a]
Block #1	Site information: Characteristics of the soil profile site, as well as its identification and classification	62
Block #2	Soil horizon description: Information on the soil morphological and other characteristics of each horizon	54
Block #3	Standard chemical analyses: Information on the standard analytical results for sampled horizons	33
Block #4	Soluble salts and heavy metals: Information on the main soluble salts and on the trace elements related to soil contamination	27
Block #5	Physical data: Information on soil physical determinations	9
Block #6	Water retention and hydraulic conductivity: Up to 25 determinations per soil sample quantifying the detailed hydraulic properties	50
Block #7	Additional analytical variables: Up to 10 specified chemical, physical, or biological characteristics	10
Block #8	Photographs: Digitized information on site, soil profile, and other plates	4
Block #9	Metadata: Information on the procedures and methods followed in preparing soil analysis data	78
Total		327

[a] Considering an average of five different horizons per soil profile, these 327 variables can generate more than 1,500 data per soil profile.

The selection of soil indicators will vary, depending upon the nature of the soil function under consideration. These soil attributes can be classified in three broad groupings: physical, chemical, or biological indicators (Table 9.3). Many of the physical and chemical soil attributes are permanent in time (inherent parameters). In contrast, biological and some physical attributes are dynamic and exceptionally sensitive to changes in soil conditions and in management practices (dynamic parameters). They appear to be very responsive to different agricultural soil conservation and management practices such as nontillage, organic amendments, and crop rotation.

The selection of soil indicator attributes should be based on: (i) land use; (ii) soil function; (iii) reliability of measurement; (iv) spatial and temporal variability; (v) sensitivity to changes in soil management; (vi) comparability in monitoring systems; and (vii) skills required for the use and interpretation (Nortcliff, 2002). As shown in Table 9.3, USDA (2006) select seven physical, three chemical, and two

Table 9.3 Soil attributes which may be used as indicators of soil quality

Grouping type	Soil indicators
Physical attributes	Soil texture[a]
	Stoniness
	Soil structure[a]
	Bulk density[a]
	Porosity
	Aggregate strength and stability[a]
	Soil crusting
	Soil compaction[a]
	Drainage
	Water retention
	Infiltration[a]
	Hydraulic conductivity
	Topsoil depth[a]
Chemical attributes	Color
	Reaction (pH)[a]
	Carbonate content
	Salinity[a]
	Sodium saturation
	Cation exchange capacity
	Plant nutrients[a]
	Toxic elements
Biological attributes	Organic matter content
	Populations of organisms[a]
	Fractions of organic matter
	Microbial biomass
	Respiration rate[a]
	Mycorrhizal associations
	Nematode communities
	Enzyme activities
	Fatty acid profiles
	Bioavailability of contaminants

[a] Key indicators selected by the USDA (2006).

biological indicators, which represent a minimal dataset to characterize soil quality. Gomez et al. (1999) define six indicators and threshold values for measuring sustainability of agricultural production systems at farm level. Other examples of soil-quality studies are reported by Doran and Jones (1996) who list soil characteristics as indicators of soil quality.

Critical limits of the soil-quality indicators are the threshold values which must be maintained for normal functioning of the soil system. Within this critical range, the soil performs its specific functions in natural ecosystems. As reported by Arshad and Martin (2002), identification of critical limits for soil-quality indicators poses several difficult problems. For example, a critical limit of a soil indicator can be ameliorated or exacerbated by limits of other soil properties and the interactions among soil-quality indicators.

For many of the soil chemical indicators, there are well-established procedures available for interpreting results. For example in chemical pollution of European soils, the Council Directive 86/278/EEC established a set of critical levels for concentration of heavy metals (Cd, Cu, Ni, Pb, Zn, Hg, and Cr). These values should not be exceeded when sewage sludge is applied in agriculture. This directive has been implemented and adapted in the form of several national or regional laws, extending the critical levels to soils in general and not limited to the application of sewage sludge. These precautionary levels are established for the cleaning up of contaminated sites, based on functional criteria and health aspects.

In the case of biological indicators, the interpretation of measurements in relation to crop yield or environmental effects is in its infancy, and there is not yet any agreed scientific basis on which to make such determinations (Wolfe, 2006). For example, increasing soil organic matter provides many benefits; however, it can also have negative environmental and crop production impacts. These negative impacts, such as requirements of many pesticides, greater P solubility, or higher soil temperature, are rarely considered or significantly weighted in soil-quality assessment (Sojka and Upchurch, 1999).

Comparing soils that have been under a certain use and management system for a number of years with natural soils that have not been disturbed, appears to be an appropriate procedure to assess soil quality by single indicators. The influence of climate, especially distribution of precipitation and temperature, geomorphology, and weathering rate could be eliminated by comparing soils exclusively within an agro-ecosystem or soil type. In this sense, it would be desirable to develop databases of the key soil indicators in natural benchmark soils with late-successional vegetation from specific ecosystems. These natural benchmark soils supporting mature vegetation would be used as the high-quality reference soils, because of the ideal balance existing between their physical, chemical, and biological properties. It would be very interesting to develop global catalogs of natural benchmark soils based on the already existing data on agricultural soils, for example, for the Mediterranean Andalusia region (Spain; De la Rosa, 1984). Obviously, these catalogs of natural soils may not be possible due to the high level of disturbance of many soils in many parts of the world.

Future technological advancements, e.g. satellite remote sensing will obviously have a positive impact on the inventory and monitoring of soil-quality indicators. The latest satellites are already covering the earth with imagery of varied spectral, temporal, and spatial resolutions. The resolutions are from 0.5 m to 24 km with days or weeks between coverage of the same area, providing a multitude of new possibilities especially through the geographical information systems (GIS) and decision support systems (DSS) implementation. The use of global positioning system (GPS) allows accurate location of observations made in the field on the inherent and dynamic indicators of the soil quality. Some scientists are using a modern technique of infrared spectroscopy (IR), in conjunction with GPS and satellite remote sensing, for rapid, nondestructive soil characterization and monitoring. From these reflectance fingerprints of soil samples can be predicted and quantified for multiple soil-quality indicators (CGIAR, 2006).

Over the last few years, the ability to extract DNA or ribosomal RNA from cells contained within soil samples, and its direct analysis in hybridization experiments has allowed to detect the presence of a vast diversity of microbes previously unimagined (Thies, 2006). Also, significant progress has been made in the development of specific biomarkers and macromolecular probes, enabling rapid and reliable measurements of soil microbial communities. Also, modern molecular biological techniques, such as fluorescence in situ hybridization (FISH) and denaturing gradient gel electrophoresis (DGGE), have facilitated the analysis of microbial biodiversity and activity; whereas the application of modern analytical techniques, such as nuclear magnetic resonance (NMR) and pyrolysis–gas chromatography–mass spectrometry (Py-GC-MS), have provided data on soil chemistry (Arias et al., 2005).

9.3.2   Soil-Quality Modeling

Within land evaluation, modeling is the fundamental component for the assessment of inherent soil quality. The models provide a tool for predicting the change in outcome caused by the changes in input parameters. By using land-evaluation models, it is possible to predict the rates and direction of many soil-quality changes. Land-evaluation modeling focuses on different purposes which can be grouped in two main classes: land suitability or productivity, and land vulnerability or degradation approaches. For example, Table 9.4 shows the MicroLEIS land-evaluation models according to the evaluated issues (De la Rosa et al., 2004).

The two principal land-evaluation modeling approaches are: (i) empirical-based modeling, and (ii) process-based modeling. The basic idea of empirical modeling for land evaluation is that observed relations are quantified and these once analyzed (i.e., in a limited number of locations) are applicable for predicting future situations. However, this will not work unless there are sufficient data on which to base the inferences, so the methodology is not appropriate for new land uses or areas from which sufficient samples have not been taken. For land evaluations of established

Table 9.4 MicroLEIS land-evaluation models according to the soil function evaluated and the concrete strategy supported for environmentally sustainable agriculture

Constituent model	Land-evaluation issue (Modeling approach)	Supported strategy
Land-use planning-related		
Terraza	Bioclimatic deficiency (Parametric)	Quantification of crop water supply and frost risk limitation
Cervatana	General land capability (Qualitative)	Segregation of best agricultural and marginal agricultural lands
Sierra	Forestry land suitability (Qualitative)	Restoration of semi-natural habitats in marginal agricultural lands: selection of forest species (22)
Almagra	Agricultural soil suitability (Qualitative)	Diversification of crop rotation in best agricultural lands: for traditional crops (12)
Albero	Agricultural soil productivity (Statistical)	Quantification of crop yield: for wheat, maize, and cotton
Raizal	Soil erosion risk (Expert system)	Identification of vulnerability areas with soil erosion problems
Marisma	Natural soil fertility (Qualitative)	Identification of areas with soil fertility problems and accommodation of fertilizer needs
Soil management related		
ImpelERO	Erosion/impact/mitigation (Neural network)	Formulation of management practices: row spacing, residues treatment, operation sequence, number of implements, and implement type
Aljarafe	Soil plasticity and soil workability (Statistical)	Identification of soil workability timing
Alcor	Subsoil compaction and soil trafficability (Statistical)	Site-adjusted soil tillage machinery: implement type, wheel load, and tire inflation
Arenal	General soil contamination (Expert system)	Rationalization of total soil input application
Pantanal	Specific soil contamination (Expert system)	Rationalization of specific soil input application: N and P fertilizers, urban wastes, and pesticides

land uses with sufficient historical or experimental data, such analyses can be very useful and are often the preferred method (Van Lanen, 1991). This empirical-based modeling has moved on from simple qualitative approaches to other procedures that are more sophisticated and based on artificial intelligence techniques.

The linking of the land characteristics with land-use requirements or limitations may be as simple as making statements about land suitability for particular uses, or lands may be grouped subjectively into a small number of classes or grades of suitability. In many qualitative approaches, quantification is achieved by the application of the rule (that is, the minimum law) that the most-limiting land quality determines the degree of land suitability or vulnerability. This assumes knowledge of optimum land conditions and of the consequences of deviations from this optimum (Verheye, 1988). Relatively simple systems of land evaluation depend largely on experience and intuitive judgment; they are really empirical models, and no quantitative expressions of either inputs or outputs are normally given. For example, the Land Capability Classification System (USDA, 1961) and its many adaptations have been widely used around the world.

Parametric methods are considered a transitional phase between qualitative methods, based entirely on expert judgment, and mathematical models. They account for interactions between the most significant factors by the multiplication or addition of single-factor indexes. Multiplicative systems assign separate ratings to each of several land characteristics, and then take the product of all factor ratings as the final rating index. These systems have the advantage that any important factor controls the rating. The most widely known method to include specific, multiplicative criteria for rating land productivity inductively was developed by Storie (1933). In the additive systems, various land characteristics are assigned numerical values according to their inferred impact on land use. These numbers are either summed, or subtracted, from a maximum rating of 100, to derive a final rating index. Additive systems have the advantage of being able to incorporate information from more land characteristics than multiplicative systems. The FAO agro-climatic zoning project represents a milestone in the development of land evaluation, introducing a new approach to land-use systems analysis (FAO, 1978).

Expert systems as computer programs that simulate the problem-solving skills of human experts in a given field have been also used. They provide solutions to a problem, expressing inferential knowledge through the use of decision trees. In land evaluation, decision trees give a clear expression of the comparison between land-use requirements and land characteristics. The expert decision trees are based on scientific background and discussions with human experts, and thereby reflect available expert knowledge. Where suitable data on practical experience are available, statistical decision-tree analysis can be used to generate land-evaluation models with good prediction rates (De la Rosa & Van Diepen, 2003).

Neural networks, as an artificial intelligence technology, have grown rapidly over the past few years and have an ability to deal with nonlinear multivariate systems. An artificial neural network is a computational mechanism that is able to acquire, represent, and compute a weighting or mapping from one multivariate space of information to another, given a set of data representing that mapping. It can identify patterns in input training data which may be missed by conventional statistical analysis. In contrast to regression models, neural networks do not require knowledge of the functional relationships between the input and the output variables. Also these techniques are nonlinear and thus may handle complex data patterns that make simulation modeling unattainable (De la Rosa et al., 1999).

The process-based models for land evaluation have been basically developed to simulate the growth of crops, along with associated phenomena that influence crop growth such as water and solute movement in soil. These simulation models are deterministic and based on an understanding of the actual mechanisms, but used to include a large empirical component in their descriptions of subsystems. The so-called Wageningen models (e.g., *WOFOST* and *CGMS*) are based on soil processes and plant physiology to predict yields under several production levels (De la Rosa & Van Diepen, 2003).

The Clouds and the Earth's Radiant Energy System (CERES) and Gamma Ray Observatory (GRO) models are probably the most widely known and used dynamic simulation models applied to agricultural production, and which are included in what is now termed Decision Support System for Agro-technology Transfer (DSSAT). The last version of these models can be parameterized to simulate several crops. DSSAT is distributed by the International Consortium for Agricultural Systems Applications (ICASA, 2006). Other dynamic simulation models apply to soil degradation aspects, such as soil erosion (e.g., EPIC, WEPP, EUROSEM), and soil contamination (e.g., LEACHM, MACRO, PEARL). Rossiter (2003) has carried out an interesting review on the application of these biophysical models in land evaluation.

The S-theory of Dexter (2004) proposes the use of an only index of soil physical quality, S. The soil physical quality index, S, is defined as the slope of the soil water retention curve at its inflection point. Examples of poor physical quality (S < 0.035) are considered by Dexter when soils exhibit one or more of the following symptoms: poor water infiltration, runoff of water from the surface, hard-setting, poor aeration, poor rootability, and poor workability. Good soil physical quality (S > 0.035) occurs when soils exhibit the opposite or the absence of the conditions listed above. S-theory appears to be also useful for predicting soil physical quality indicators, for example, hydraulic conductivity, friability, compaction, penetrometer resistance, and root growth.

A major impediment to applying process-based models in soil-quality evaluation is the requirement for high-quality and high-frequency data of soil indicators. However, missing soil indicators can be estimated with pedotransfer functions from routine soil survey data, although these approximations will lead to less successful applications (Pachepsky & Rawls, 2004). Also, the combination of dynamic simulation models and empirically based land-evaluation techniques are currently producing good scientific and practical results, improving the accuracy and applicability of the models.

9.3.3   Integrated Approach

For integrated soil-quality assessment, the development of relationships between all the soil-quality indicators and the numerous soil functions may be a monumental task (Zalidis et al., 2002). Therefore, a stepwise agroecological approach for soil-quality evaluation and monitoring was proposed by De la Rosa (2005). Two steps relating to: (i) inherent soil quality, and (ii) dynamic soil quality are involved (Fig. 9.2).

Step #1. Land evaluation is an appropriate procedure for analyzing inherent soil quality from the point of view of long-term agroecological changes. Within this complex context, land-evaluation models may serve as a first step to develop a soil-quality assessment procedure (Arshad & Martin, 2002). The first step will result in defining agroecological zones, land suitability, and vulnerability classes, for example, by application of MicroLEIS models (De la Rosa et al., 2004).

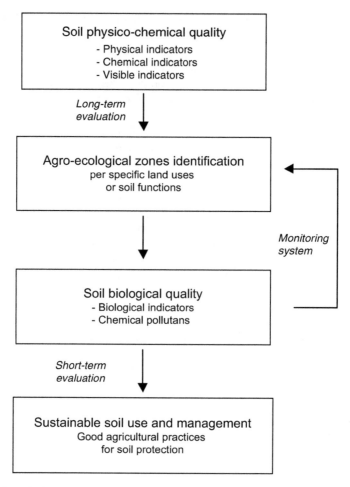

Fig. 9.2 Graphical representation of a stepwise agroecological approach for soil-quality assessment

Step #2. A short-term evaluation and monitoring procedure would be basically considered for the soil biological quality in each agroecological zone defined in the first step. By measuring appropriate indicators, changes in soil dynamic quality can be assessed. These indicators would be compared with the desired values (critical limits or threshold level), at different time intervals (Arshad & Martin, 2002). This comparison of single indicators should be of natural soils that have not been disturbed with soils that have been under a certain use and management system for a number of years. Because soil biological parameters are most variable and sensitive to management practices, a monitoring system (observed change over time) would provide information on the effectiveness of the selected farming system, land-use practices, technologies, and policies. For example, dehydrogenase activity in Mediterranean forest soils proved to be very sensitive to both natural and management changes, and showed a quick response to the induced changes (Quilchano & Marañon, 2002). Also, enzyme activities have been found to be very responsive to different agricultural management practices such as nontillage (Bergstrom et al., 1998).

Because of the complex nature of the soil and its high spatial and temporal variability, it is appropriate to develop soil-quality assessment based on biological indicators after the traditional land evaluation using basically physicochemical parameters. This agroecological approach should focus on dynamic soil aspects (biological factors) but with awareness of inherent soil aspects (physical and chemical factors).

In the ensuing section, we present a case study to illustrate the assessment of the inherent and dynamic aspects of soil quality. Land-evaluation models are used to express inherent soil quality, whereas dynamic soil quality is evaluated by assessing the effects of land use and management on soils through monitoring soil properties that more readily respond to use and management.

9.3.4   A Case Study: Assessing Soil Quality in Argentina

Background. The Pampean Region covers about 560,000 km², and includes the provinces of Buenos Aires, Santa Fe, Entre Rios, Cordoba, and La Pampa in Argentina. It has a semihumid to humid subtropical climate with annual precipitation ranging from 800 to 1200 mm. Large interannual variability in rainfall characterizes the region. Mean annual temperature of around 16°C occurs in July, the coldest month, whereas January is the warmest month. Altitude varies from 10 to 750 m with a mean of about 50 m, and slope gradients vary from flat to more than 5%. Formerly, grazing was an important activity, but at present rainfed agriculture is the main economic activity in the region. Conversion of grassland into cropland is the major land-cover process during the last 10 years, accounting for about 28% of increase of cultivated land area. The major agricultural crops are wheat, maize, soybean, and sunflower (Moscatelli & Sobral, 2005). The formation of pampean soils is influenced by a flux of eolian materials which originate from unstable desert surfaces. The "pampean loess" are primarily vitric in nature (volcanic glass) and rich in calcium carbonate. Since 2003, the INTA Soil Institute had developed the inventory and monitoring of selected benchmark soil series from Pampa Region, for undisturbed and cultivated sites. This project is based on the previous soil survey studies in the region, where the largest soil map with the largest spatial extent was produced at a scale of 1:500,000 though a major part of the area was also mapped at 1:50,000.

For this case study, the selected benchmark soil is Ramallo series (Figs. 9.3 and 9.4), covering about 232,000 ha in the high plains near to Parana River, in the northwest of Buenos Aires province. The USDA taxonomic classification of a typical pedon of Ramallo series is Fine-loamy, Mixed, Thermic, Vertic Argiudolls, with soil horizons A, Bt, C of a dark grayish brown color, silty clay loam texture, very deep and somewhat poorly drained (Location: 33° 40′ 50′ S and 60° 03′ 10′ W).

Since 1997, the dominant farming system on Ramallo soils is the direct seeding on permanent soil cover (DSPSC), in wheat–soybean rotation (Fig. 9.5). Table 9.5 shows the typical operations sequence corresponding to this farming system for each crop. The DSPSC system is different from conventional agriculture

Fig. 9.3 Topsoil of Ramallo series soil, under natural conditions. The A1 horizon has a depth of 31 cm, and an organic matter content of 4.27%

Fig. 9.4 Topsoil of Ramallo series soil, under cultivated conditions. The Ap horizon presents a depth of 18 cm, and an organic matter content of 3.10%

Fig. 9.5 Plot cultivated with direct seeding on permanent soil cover (DSPSC method), soybean crop over wheat stubble on Ramallo soil. The soil cover is not incorporated into the soil by tillage

Table 9.5 Operation sequence of direct seeding on permanent soil cover (DSPSC farming system) on the Ramallo soil in the Pampa region, Argentina

Operation	Time (month)	Implement type	Material input	Work rate (h/ha)
Wheat crop				
Weed control	June	Motor sprayer	Glyphosate (2 l/ha) + 2-4D (0.5 l/ha)	0.40
Crop planting	June	Direct seeder drill	Seed (130 kg/ha)	1.00
Fertilizer application	June	Direct seeder drill	Phosfate ammonia (70 kg/ha) + Urea (80 kg/ha)	
Weed control	August	Motor sprayer	Metsulphoron (5 g/ha)	0.40
Harvest	December	Combine		0.75
Soybean crop				
Weed control	September	Motor sprayer	Glyphosate (2 l/ha) + 2–4D (0.5 l/ha)	0.40
Weed control	October	Motor sprayer	Glyphosate (2 l/ha) + 2-4D (0.5 l/ha)	0.40
Crop planting	October	Direct seeder drill	Seed (80 kg/ha)	1.00
Fertilise	October	Direct seeder drill	Phosphate mono ammonia (60 kg/ha)	
Weed control	November	Motor sprayer	Glyphosate (2 l/ha)	0.40
Pest control	December	Motor sprayer	Monocrotophos (0.6 l/ha)	0.40
Pest control	January	Motor spray	Endosulfan (0.6 l/ha)	0.40
Harvest	April	Combine		0.75

Table 9.6 Comparison of average values[a] of soil-quality indicators corresponding to the Ramallo benchmark soil, for natural and cultivated conditions, in the Pampa Region, Argentina

Soil quality Indicator	Natural conditions (n = 6)	Cultivated conditions (n = 20)	Change intensity, %
Inherent soil quality			
Color (humid)	10YR3/2	10YR4/2	
Textural class	Clay silty loam	Clay silty loam	
Clay content	56.5	55.4	2
Cation exchange capacity	26.5	22.3	16
Reaction, pH	7.5	6.8	10
Bulk density (g cm^{-3})	1.07	1.13	5
Dynamic soil quality			
Organic matter content (%)	4.3	3.1	22
Respiration rate (kg C ha^{-1} day^{-1})	83	61	27
Topsoil loss (%)	0	31	31
Aggregate stability (%)	70	59	16
Infiltration (mm h^{-1})	44	20	55
Compaction (Mpa)	3.7	4.9	32
Structure index (%)	80	53	34

[a]The average values for natural and cultivated conditions correspond to 6 and 20 different sites, respectively. Sampling period: 2003–2005, at planting time.

in that it retains crop residues on the soil surface as a cover, not incorporating them into the soil by tillage. Crop residues are used to form suitable mulch that protects the soil and suppresses weed growth. However, this ground-cover strategy requires herbicides for weed control, especially for the soybean crop (Table 9.5). Therefore, this no-tillage system is characterized by a very high dependency on chemical external inputs. The conventional farming system previously used on these soils included a very intensive soil tillage with several chisel, disk plough, and tine operations.

Step #1: Inherent soil quality. The single indicators of inherent soil quality presented in Table 9.6 are relatively permanent, and the difference between natural and cultivated conditions is relatively low (2–16%). According to different agroecological land-evaluation models, the soil and climate information of Ramallo Series have been interpreted as follows.

- USDA (1961) Land Capability Classification: Class II **e**—Very good soils with few limitations due to soil erosion that reduce the choice of crops or require some conservation practices. This land-evaluation system considers eight capability classes (I to VIII) with a decreasing production potential in terms of expected yield and the range of crops that can be grown. The subclass **e** represents erosion hazard.

- FAO (1978) agro-climatic evaluation: 270 < GPL < 330 days; where GPL is the length of growing period calculated on the basis of the annual precipitation (800–1,200 mm), the annual potential evapotranspiration (1,100 mm), and the available stored soil moisture (100 mm). This represents very high biophysical crop production potential under rainfed conditions.
- The Universal Soil Loss Equation (USLE; Wischmeier and Smith, 1965) erosion risk evaluation: Estimated current soil loss is 20.5 t ha^{-1} year^{-1}. The USLE prescribes 10 t ha^{-1} year^{-1} as acceptable level of soil erosion. Therefore, Ramallo soils require some conservation practices.

Step #2: Dynamic soil quality. The dynamic aspect of soil quality is the focal point for maintaining soil health. Dynamic quality results from the changing nature of soil properties that are influenced by human use and management decisions. Collectively, the effects of management will either result in a net positive or negative impact on the quality of the soil. For Ramallo soils, Table 9.6 shows the values measured for some dynamic indicators at natural or undisturbed and cultivated sites. The record of management practices (DSPSC farming system) followed in the cultivated sites is summarized in Table 9.5. The last column of Table 9.6 suggests that several degradation processes are in progress with different intensity. These are mainly subsoil compaction (decrease in infiltration by 55%, and increase in compaction by 32%); and water erosion (decrease in topsoil depth by 31%, and organic matter content by 22%). Soil contamination by herbicides, which has not been studied on the cultivated soils may be another important degradation process.

9.4 Soil-Quality Assessment Implementation

The agroecological land-evaluation support system MicroLEIS (De la Rosa et al., 2004) can be quoted as an example in the application or generalization phase of soil-quality assessment approaches. This phase will make possible the practical use of the information and knowledge gained during the prior phase of developing assessment procedures. The MicroLEIS system was developed to assist specific types of decision-makers faced with specific agroecological problems in the Mediterranean region. It has been designed as a knowledge-based approach which incorporates a set of information tools, as illustrated in Fig. 9.6. Each of these computer-assisted procedures is directly linked to others, and customized applications can be carried out on a wide range of problems related to land productivity and land degradation.

9.4.1 Data Warehousing

Data warehousing can be greatly facilitated if the nearly infinite list of basic attributes is systematically arranged and stored in an ordered format for ready

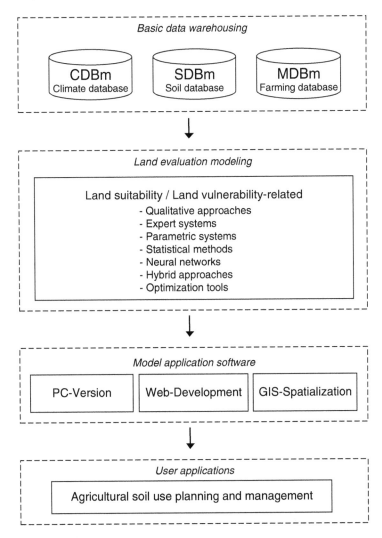

Fig. 9.6 Conceptual design and component integration of the MicroLEIS DSS land-evaluation decision support system. (Adapted from De la Rosa et al., 2004. With permission)

sorting and retrieval. Database management systems are responsible for these tasks and consist of attribute tables manipulated by relational database management systems, and a geometric component handled by GIS.

The land attributes used in MicroLEIS correspond to the following three main factors: soil/site, climate, and crop/management. Soil surveys are the building blocks of the comprehensive data set needed to drive land evaluation. Land evaluation is normally based on morphological, physical, and chemical data derived from the soil survey, such as soil depth, texture, water capacity, drainage class, soil reaction, and organic matter content. Other biophysical factors, mainly referred to monthly climate parameters, are also considered land characteristics. Because climatic conditions vary from year to year, reliable long-term data are used to reflect the historical reality and to predict future events with some degree of confidence. Traditionally, agricultural management aspects have been considered a prerequisite only in land evaluation. Today, management factors are being incorporated as input variables in response to a growing need for integrating farming information. In this sense, crop and management data derived from field observation, monitoring, or experimentation, such as growing-season length, rooting depth, tillage operations, and treatment of residues, are also considered land characteristics.

For each of these main factors, a relational database has been constructed: SDBm Plus, CDBm, and MDBm, with connections between the three databases. The multilingual soil database SDBm Plus (De la Rosa et al., 2002) is a geo-referenced soil attribute database for storage of an exceptionally large number of morphological, physical, and chemical soil profile data. This database is the "engine" of the MicroLEIS DSS system. It is a user-friendly software designed to store and retrieve efficiently and systematically the geo-referenced soil attribute data collected in soil surveys and laboratories. As illustrated in Fig. 9.7, the SDBm Plus database has the following main characteristics: (i) running on WINDOWS platforms; (ii) "help menus" facilitating data entry; (iii) automatic translation from English to Spanish, French, and German; (iv) metadata feature to describe the methods used in laboratory analysis; (v) temporal mode to collect over time physical and hydraulic soil properties; (vi) structured query procedure to allow detailed searches; (vii) simple graphical analyses and report generation; and (viii) an input file generator for the automatic transfer of the stored soil attribute data to GIS and computerized land-evaluation models. The SDBm Plus database was developed by the MicroLEIS Group with the collaboration of FAO through a joint project (FAO-CSIC, 2003).

The climate database CDBm developed for MicroLEIS DSS is a computer-based tool for the organization, storage, and manipulation of agro-climatic data for land evaluation. These geo-referenced climate observations from a particular meteorological station correspond to the mean values of such records for a determinate period. The basic data of CDBm are the mean values of the daily dataset for a particular month. The stored mean monthly values correspond to a set of temperature and precipitation variables (maximum temperature, minimum temperature, accumulative precipitation, maximum precipitation per day, and days of precipitation).

The farming database MDBm is knowledge-based software to capture, store, process, and transfer agricultural crop and management information obtained through interviews with farmers. Each MDBm dataset consists of geo-referenced agricultural information on a particular land-use system. This structured collection of information is stored as a database file. A menu system guides the user through a sequence of options to capture the management practices followed on a site-specific farm. Input parameters are farm and plot descriptions, crop characteristics, sequence of operations, and behavioral observations. These parameters represent a total of 59 default variables according to good management practices on Mediterranean farms.

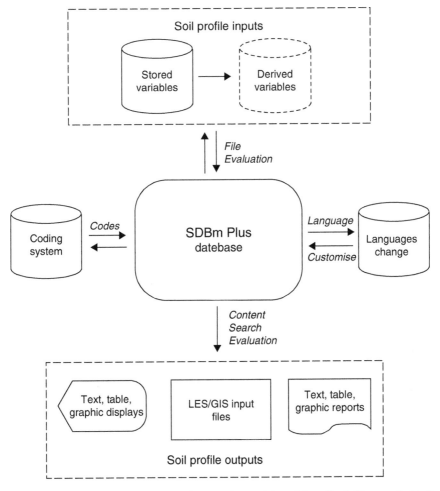

Fig. 9.7 General scheme of the SDBm Plus database. (Adapted from De la Rosa et al., 2002. With permission)

9.4.2 Modeling Integration

The possibilities for using land-evaluation models in decision-making by developing the model application software are enormous. This integration phase will make possible the practical use of the information and knowledge gained during the prior phase of building evaluation models (Antoine, 1994). When the assessment models are expressed in notations that can be understood by a calculating device, the algorithms become computer programs. In order to put the models to use in practical applications, that is, to automate the application of land-evaluation models, a library of PC-based software is developed. A graphical interface is designed which will allow the models to be easily applied. This user interface is considered a very important component because, to the user, it *is* the system.

Within the MicroLEIS framework, the PC-based software has been written using various programming languages, particularly Basic and C++. It has the following main characteristics: (i) input data through the keyboard and connection with the attribute databases; (ii) "pop up" screens showing codes, types, and classes of input variables; (iii) models running in individual and batch-processing modes; (iv) output evaluation results in window, printout, and file formats; and (v) links of output files with GIS databases. These computer programs are largely self-explanatory.

The model computer programs can also be implemented on the Internet through a World Wide Web (WWW) server, so that users can apply the models directly via a Web browser. It is not necessary to download and install the PC software. These open-access WWW applications offer several advantages, such as their use by many people, allowing their usability to be checked in order to improve the systems. Upgrades are immediately made available on the WWW server. The Web site is the center of activity in developing operative planning or decision support systems.

9.4.3 Application Tools

Spatial decision support systems for policymakers and land users must focus on choosing optimal use and management decisions. In this sense, optimization tools based on land-evaluation models are very important in formulating decision alternatives—for example, agricultural management practices to minimize threats to the sustainability of farming systems. Agricultural management operations depending on spatially varying land characteristics have the added difficulty of trying to satisfy multiple, and often opposing, aims: the best soil conditions for plant growth may not be the best with regard to erosion or pollution.

Within the MicroLEIS, the optimization tools are used in conjunction with running various models. On the basis of the quadratic version of the *Albero* model, a mathematical procedure was developed to determine a combination of input variables to maximize predicted yields. This procedure involved taking the first mathematical derivative with respect to each independent variable, setting it to zero, and solving the system of simultaneous equations (De la Rosa et al., 1992). On the basis of the expert-system/neural-network structure of the *ImpelERO* model, a computerized procedure was followed to find an appropriate combination of management practices to minimize soil loss for a particular site (specified climate and soil characteristics). This formulation of specific crop management for soil protection of each particular site is one of the most interesting features of the *ImpelERO* model (De la Rosa et al., 2000).

Spatial or regional analysis includes the use of spatial techniques to expand land-evaluation results from point to geographic areas, using soil survey and other related maps. The use of GIS technology leads to the rapid generation of thematic maps and area estimates, and enables many of the analytical and visualization operations to be carried out in a spatial format, by combining different sets of information in various ways to produce overlays and interpreted maps. Furthermore, digital satellite images can be incorporated directly into many GIS packages. This technology is a prerequisite for managing the massive datasets required for spatial land-evaluation application—a simple map subsystem (e.g., ArcView) being all that is required to show basic data and model results on a map, or to extract information from maps to be used in the land-evaluation models.

The option "Spatialization" of MicroLEIS was developed as a further stage of the scaling-up process of evaluation models application. GIS technology was used to extract information from maps to be used in the predictive models, and to show model results on a map. The evaluation results are estimated by grid cells and aggregated to regional level. The soil survey maps, which in geographical format are usually polygonal multifactor maps, are the main source of basic information. Additional basic information can be extracted from other soil-survey-related maps, such as land-use maps. At the regional scale, part of the basic information for applying MicroLEIS land-evaluation methods can be facilitated by single-factor grid maps, such as digital terrain models, along with satellite images (De la Rosa et al., 2004).

9.5 Sustainable Management Practices

Rule number one is that universal recommendations on sustainability of soil management practices must not be done. It is clear that each particular site (combination of climate, soil type, and land use) requires a different set of management practices, but several general principles can apply in most situations. These general principles on sustainable agricultural practices focus on the positive effects on soil quality: (i) increased organic matter, (ii) decreased erosion, (iii) better water infiltration, (iv) more water-holding capacity, (v) less subsoil compaction, and (vi) less leaching of agrochemicals to groundwater. All these soil-quality conditions are essential for the proper functioning of soil, and one or two of them alone will not be enough. They can be analyzed in relation to the following groups: (i) arable land identification, (ii) crop diversification, (iii) organic matter restoration, (iv) tillage intensity, and (v) soil input rationalization.

9.5.1 Arable Land Identification

Agricultural management systems located on the most suitable arable lands, according to their agroecological potentialities and limitations, is the first step to achieve soil sustainability. On the contrary, any kind of agricultural management system will have a negative environmental impact when applied on land with very low suitability for agricultural uses. Marginal agricultural land under any kind of farming system used to be the ideal scenario for soil erosion. Therefore, a positive correlation between current land use and potential land capability would be desirable, beginning with the identification of the best agricultural lands.

It is clear that in many marginal agricultural lands, it can be necessary to change the land-use system fundamentally by conversion from arable to forest or pasture. For example, the case of a Mediterranean region: Andalusia, as shown in Table 9.7 the relationship between present land use (current use) and agricultural land capability (potential use) is clearly unbalanced (De la Rosa & Moreira, 1987). About 1 million ha of rainfed agricultural lands must be converted to forestry, grazing, or natural lands in order to get a better equilibrium in comparison with the moderately or clearly marginal lands. Similar situations are very frequent in other European regions, and it is the major reason for the reforestation programs launched by the European Commission.

9.5.2   Crop Diversification

Crop diversity is beneficial for several reasons. Each crop contributes a unique root structure and type of residue to the soil. A diversity of organisms can help control pest populations, and a diversity of cultural practices can reduce weed and disease pressures. Diversity across the landscape and over time can be increased by using buffer strips, small fields, contour strip cropping, crop rotations, and by varying tillage practices. Changing vegetation across the landscape or over time increases plant diversity, and the types of insects, microorganisms, and wildlife (USDA, 2006). In contrast, simplification of crop rotation as a relevant element of arable intensification has led to soil deterioration and other negative environmental impacts.

Table 9.7 Comparison of present land uses and agroecological land-capability classes in Andalucia region, Spain

Category	Estimated extension (10^3 ha)	Percentage (%)
Irrigated agricultural lands	592	7
Present land use		
Rainfed agricultural lands	3,165	36
Forestry, grazing and natural lands	4,007	46
Others	936	11
Land capability class		
S1. Excellent agricultural lands	535	6
S2. Good agricultural lands	1,735	20
S3. Marginal agricultural lands	2,311	27
N Nonagricultural lands	4,073	47

Within the agricultural lands, all soils can be used for almost all crops if sufficient inputs are supplied. The application of inputs can be such that it dominates the conditions in which crops are grown, such as it can be the case in greenhouse cultivation. However, each soil unit has its own potentialities and limitations (soil suitability), and each crop its biophysical requirements. In order to minimize the socioeconomic and environmental costs of such inputs, the second major objective in managing soil quality is to predict the inherent suitability of a soil unit to support a specific crop for a long period of time. This kind of study provides a rational basis to diversify agricultural soil system considering all the possible crops (De la Rosa & Van Diepen, 2003).

9.5.3   Organic Matter Restoration

Increasing soil organic matter level is critical because organic matter is related to many aspects of soil physical, chemical, and biological quality. Organic matter improves soil structure, water-holding capacity, nutrient availability, biological activity, and can help protect against erosion and compaction. Also, soil organic matter restoration could remove significant amounts of carbon dioxide from the atmosphere (see Chapter 2).

Better crop yields that produce more crop residues to incorporate into the soil are the best way to increase stable soil organic matter. Burning the straw or stubble after harvest is a practice not recommended. Further additions may come from animal manure, green manure, and sewage sludge and biosolid wastes when properly and safely recycled. To do the last, it is necessary to select the more appropriate kind of soil to receive these wastes. For example, calcareous soils appear to be the most suitable, considering the important role of calcium for increasing the efficiency of accumulation of soil organic matter. Also, soil contamination vulnerability, specially referred to heavy metals, must be considered in selecting appropriate application sites.

In general terms, it has been estimated that an annual return of $5\,t\,ha^{-1}$ of crop residues could keep soils in equilibrium with present levels of soil organic matter (Wallace, 1994). In the Mediterranean region, where it is hotter, some more tons per ha will be needed. The efficiency of conversion of such carbon to stable soil organic matter is not constant and is a function of several variables.

9.5.4   Tillage Intensity

Soil tillage has positive agricultural effects, preparing suitable seedbeds for crops; controlling weeds; and incorporating manure, fertilizers, pesticides, and other amendments. At the same time, the negative consequences of tillage practices accelerate soil erosion and compaction process, by destroying soil organic matter and soil structure. To formulate the tillage type and intensity for each particular soil is critical for tackling soil degradation problem in agricultural lands (De la Rosa et al., 1999).

In general terms, tillage systems can range from full-width intensive tillage to zero tillage (i.e., intensive tillage, reduced tillage, ploughless tillage, minimum tillage, and no-tillage). The most common intensive tillage system of dry farming consists of moldboard ploughing to break the hardened soil surface, and many successive disking and harrowing to reduce soil-clod size. Traditional tillage implements (e.g., plow moldboard or disk cultivator) which cause soil inversion can be especially appropriate for high-slope soils due to increase surface roughness (>30 mm). Increasing the surface roughness (micro-topography) along the contour direction decreases the transport capacity and runoff detachment by reducing the flow velocity. During a rainfall event, rough surfaces are eroded at lower rates than smooth surfaces under similar conditions. The soil workability status for each soil and tillage operation is very related with the produced surface roughness and other soil physical properties. The soil workability status ("tempero" in Spanish language) is considered as the optimum soil water content where the tillage operation has the desired effect of producing the greatest proportion of small aggregates (Dexter & Bird, 2001). Beyond this soil water range, soil is too wet or too dry, and therefore the tillage operation alters in an adverse way the soil physical properties and facilitates soil erosion. By taking into account the water workability limits for each soil and tillage operation, it is possible to reduce the effects of soil erosion. Soil workability and its influence on soil tillage is widely analyzed by Dexter (2004) as an important aspect of his interesting S-theory on soil physical quality. Subsoiling, deep ploughing, para-ploughing, and numerous other tillage implements can be used to alleviate the problems created by subsoil compaction.

However, this conventional repeated tillage system accelerates decomposition of organic matter thus affecting soil physical, chemical, and biological attributes of soil quality (Moreno et al., 2006). Topsoil pulverization by repeated tillage and under dry soil conditions has a very negative effect on erosion. Finely pulverized soils are usually smooth, seal rapidly, and have low infiltration rates, as might be the case for some roto-tilling operations or for repeated cultivations of silt loam soils under Mediterranean conditions. The subsoil compaction caused by tillage and traffic with increasing weight of agricultural machinery is a problem especially severe in heavy-textured and poorly drained soils. Increased soil-bulk density reduces air permeability, water infiltration, and sometimes root development. An intensive tillage system is clearly inappropriate for most soils and must be avoided to minimize soil erosion.

With the no-tillage system or DSPSC (also named conservation agriculture) the soil is left undisturbed, including direct sowing and weeds control accomplished with herbicides. Although there are several forms of conservation agriculture, normally it is considered synonymous with no-tillage systems. The DSPSC system is gaining popularity among farmers from South America and USA. In 2002, Derpsch and Benites (2003) calculated that the total world area covered was 72 million ha, with 46% corresponding to South America and 31% to USA. There is overwhelming evidence from several scientific studies that continuous no-tillage in some agro-ecosystems is the most effective and practical approach for restoring and improving soil quality (Arshad, 1999). With successful conservation agriculture, the loss of soil organic matter can be very much reversed. The erosion risk decreases as the soil surface is continuously covered, mainly during the rainy season. Conservation agriculture, which provides a high level of crop residues, increases the biodiversity level—more wildlife and soil fauna. In Mediterranean agro-ecosystems, the best results of no-tillage system seem to be obtained on the heaviest clay soils (Gomez et al., 1999).

However, the level of success of this agriculture varies with (i) site and climatic conditions, (ii) soil type, (iii) crop species, and (iv) growing period length. So, high slope gradient (>15%) appears to be a limiting factor to introducing conservation agriculture (Martinez-Raya, 2003). With special reference to the Mediterranean region, in soils with low water infiltration rate and prone to surface sealing the effects of no-tillage can increase runoff generation and erosion problem (Gomez et al., 1999). A short growing-period length (GPL < 250 days; e.g., in Scandinavian or Mediterranean agro-ecosystems) is considered a barrier to adoption of no-tillage system, due to stunted development of the mulching horizon (Arshad, 1999). The GPL is one of the major parameters in agro-ecological zoning studies (land-use planning). The increased density of the soil just beneath the depth of tillage (subsoil compaction) is one of the most striking effects of management system, specially plough-less tillage. Because in many cases the mulch cover is not sufficient to suppress weed growth, it is needed to use herbicides to control weeds, increasing the soil-contamination risk by leaching of agrochemical to groundwater.

In summary, the general trend in soil tillage system would be to (i) reduce tillage intensity, (ii) follow the contour for tillage direction, (iii) diversify tillage implements, (iv) reduce subsoil compaction, and (v) consider optimum soil workability.

9.5.5 Soil Input Rationalization

Over the last four decades, it is evident that chemical applications have revolutionized agriculture. On the positive side, fertilizers and pesticides have increased crop production and the amount of organic matter returned to the soil. However, soil and water contamination is very high in many places with increased agricultural intensification. Independently of the nutrient needs for crop yield, the application of fertilizers usually exceeds the functional capacity of the soil to retain and transform such nutrients. In many cases, the saturation of the soil with nitrogen and phosphate has led to losses of nitrates into shallow groundwater and saturation of the soil with phosphate, which may also move into the groundwater (Zalidis et al., 2002).

The risk of applying manure and urban wastes (basically sewage sludge and compost) on agricultural soil must be considered based on three components relevant to soil protection: organic matter content, nutrient load, and contaminant load.

The maximum risk from the extensive use of pesticides is due to leaching and drainage of pesticides into the surface- and groundwater. Several soil functions can be degraded, including the food web support, the retention and transformation of toxicants and nutrients, and soil resilience. The frequent use of herbicides is drastically changing the methods of crop production, but their impacts on soil quality/degradation are still not known exactly. The exclusive chemical weed control must be identified as an important limiting factor in the adoption of the no-tillage system. In this case, the risk of soil contamination by herbicides must be analyzed because, ironically, farming practices to remedy eroded soils can increase soil degradation by contamination.

9.5.6 Innovative Examples

In the decades ahead, the development of sustainable agricultural systems will require great improvements, not just through biotechnology and chemical use but also via agro-ecological innovations (Uphoff, 2002). As referred by some authors, in the future "a doubly green revolution" will be necessary that reverses environmental deterioration at the same time that it augments the supply of food. This section makes reference to a variety of innovative agro-ecological methods, basically interventions that target biological processes, already used to have a positive impact on soil quality and crop production: (i) agroforestry interventions, such as the addition of "tithonia" as a green manure; (ii) conservation agriculture with no-tillage, now evolving into comprehensive soil management strategies; (iii) small farm management, by a combination of terracing, windrows, applications of manure, water harvesting, etc.; (iv) bio-intensive agriculture, including double-dug beds, optimal spacing of plants, use of organic nutrients, and other "permaculture" techniques; (v) soil bio-rehabilitation methods, introducing new vegetation that supports intensified microbial interactions, rhizobia, mycorrhizal, trichoderma, etc., as a kind of microbiological weathering of the soil; (vi) managed fallows to fertility recovery, by planting and managing certain plants, for example, *Crotalaria* and *Chromaleana*, in fallow rotation systems; (vii) green manure and cover crops, to restore degraded soils or raise productivity of cropped soils; (viii) modification of soil horizons, breaking up lower soil horizons and aerating them to a depth of several feet; (ix) composting and vermicomposting, combining worm action and composting to change soil organic matter, both in quantity and in quality; (x) slash-and-char cultivation, incorporating incompletely burned charcoal in the soil; (xi) polycropping management, to control plant pests through selected mixes of crops; and (xii) biological remediation of pesticides, heavy metals, and other contaminants, by microorganisms genetically engineered to break down or take up contaminants. Uphoff et al. (2006) presents an interesting "state-of-practice" review on these biological strategies for a new agriculture. The book provides a comprehensive understanding of the science and steps needed to utilize soil systems for sustainable agriculture.

9.6 Conclusion

Maintenance and improvement of soil quality is one of the most important prerequisites to achieve the environmental sustainability. In spite of the huge controversy (Sojka & Upchurch, 1999), the modern concept of soil quality is a valid and important framework in interpreting scientific soil information and predicting sustainable soil use and management. However, the process of evaluating soil is not new, and agro-ecological land evaluation, developed since the middle of the twentieth century (Davidson, 1992), has much to offer in the complex task of soil-quality assessment.

Soil-quality indicators are valuable tools and are finding increasing application. However, dynamic soil indicators should be measured after estimation of inherent soil indicators. An agroecological approach follows two steps: (i) developing long-term, inherent, specifically physicochemical evaluation, and (ii) short-term, dynamic, specifically biological evaluation (De la Rosa, 2005). The focus on biological approaches must not diminish appreciation of the physical and chemical factors, in order to develop a productive integration of the three sets of factors. The selected case study points out that soil-quality aspects are inherent and dynamic, and that can be measured and explained through land-evaluation modeling and simple indicators comparison, respectively. The analyzed DSPSC system can be considered a mulch-based management system adapted to the particular condition of Ramallo soils in Argentina, although with the major disadvantages of high dependence on chemical inputs (mainly herbicides) and limited crop diversification because of the need of cereals in the crop rotation.

Modern technologies are providing unprecedented power and flexibility plus the possibilities to combine soil-quality information and knowledge in novel and productive ways. The agroecological land-evaluation decision support system such as MicroLEIS reflects the many advances in these technologies and their possibilities for the development and application to soil-quality assessment (De la Rosa et al., 2004).

Sustainable soil management can maintain and even improve soil quality through the use of soil-specific practices, adapted to local soil, terrain, and climatic conditions, by using decision or planning support tools. The agro-ecological paradigm for a new agriculture defended in this chapter needs to be considered under two central perspectives: site specificity and time dimension. However, several general principles can apply in most situations across international boundaries. These basic principles on sustainable agricultural practices focus on the positive effects on the soil quality: (i) increased organic matter, (ii) decreased erosion, (iii) better water infiltration, (iv) more water-holding capacity, (v) less subsoil compaction, and (vi) less leaching of agro-chemicals to groundwater. To achieve these objectives, the following sustainable soil use and management strategies will be developed: (i) arable land identification, (ii) crop diversification, (iii) biomass restoration, (iv) appropriate tillage intensity, and (v) soil input rationalization.

In the future, a postmodern agriculture has potential for great improvement, not just through biotechnology and the use of chemicals but also via agroecological innovations in order to increase the crop production and environmental protection. This will depend crucially on soil quality and the methods for its assessment; being an area of knowledge generation and practical application where the science is still young. It can be anticipated that farmers from different geographical contexts will begin to want to apply information technology to support many soil-specific operational aspects of farming in the future, for example, real-time decision support systems (Thysen, 2000).

Acknowledgments The authors would like to thank Prof. D. Davidson, University of Stirling (Scotland, UK), for his constructive comments on this chapter and suggestions for modifications. Thanks are also expressed to Ing. Roberto Casas, Director INTA Soil Institute in Argentina, for his collaboration in developing the case study.

References

Antoine, J. (1994). Linking geographical information systems (GIS) and FAO's agro-ecological zone (AEZ) models for land resource appraisal. *FAO World Soil Resources*, Report 75. Rome, FAO.

Arias, M. E., Gonzalez-Perez, J. A., Gonzalez-Vila, F. J., & Ball A. (2005). Soil health-a new challenge for microbiologists and chemists. *International Microbiology*, 8, 13–21.

Arshad, M. A. (1999). Tillage and soil quality: Tillage practices for sustainable agriculture and environmental quality in different agro-ecosystems. *Soil & Tillage Research*, 53, 1–2.

Arshad, M. A., & Martin, S. (2002). Identifying critical limits for soil quality indicators in agro-ecosystems. *Agriculture, Ecosystems & Environment*, 88, 153–160.

Ball, A., & De la Rosa, D. (2006). Modeling possibilities for the assessment of soil systems. In: N. Uphoff, A. Ball, E. Fernandes, H. Herren, O. Husson, M. Laing, Ch. Palm, J. Pretty, P. Sanchez, N. Sanginga, & J. Thies (Eds.), *Biological approaches to sustainable soil systems* (pp. 683–692). Boca Raton, FL: Taylor & Francis/CRC Press.

Bergstrom, D. W., Monreal, C. M., & King, D. J. (1998). Sensitivity of soil enzyme activity to conservation practices. *Soil Science Society of America Journal*, 62, 1286–1295.

CGIAR. (2006). Consultative Group on International Agricultural Research. Available at http://www.cgiar.org. (Retrieved on 24 February 2006.)

Davidson, D. (1992). *The evaluation of land resources*. Harlow, Essex, UK: Longman.

Davidson, D. (2000). Soil quality assessment: Recent advances and controversies. *Progress in Environmental Science*, 2, 342–350.

De la Rosa, D. (Ed.) (1984). *Catalogo de suelos de Andalucia*. Sevilla, Spain: Agencia de Medio Ambiente, Junta de Andalucia.

De la Rosa, D. (2005). Soil quality evaluation and monitoring based on land evaluation. *Land Degradation & Development*, 16, 551–559.

De la Rosa, D., & Moreira, J. M. (Eds.) (1987). *Evaluacion agro-ecologica de recursos naturales de Andalucia*. Sevilla, Spain: Agencia de Medio Ambiente, Junta de Andalucia.

De la Rosa, D., Moreno, J. A., Garcia, L. V., & Almorza, J. (1992). MicroLEIS: A microcomputer-based Mediterranean land evaluation information system. *Soil Use & Management*, 8, 89–96.

De la Rosa, D., Mayol, F., Moreno, J. A., Bonson, T., & Lozano, S. (1999). An expert system/neural network model (ImpelERO) for evaluating agricultural soil erosion in Andalucia region. *Agriculture, Ecosystems & Environment*, 73, 211–226.

De la Rosa, D., Moreno, J. A., Mayol, F., & Bonson, T. (2000). Assessment of soil erosion vulnerability in western Europe and potential impact on crop productivity due to loss of soil depth using the ImpelERO model. *Agriculture, Ecosystems & Environment, 81*, 179–190.

De la Rosa, D., Mayol, F., Moreno, F., Cabrera, F., Diaz-Pereira, E., & Antoine, J. (2002). A multilingual soil profile database (SDBm Plus) as an essential part of land resources information systems. *Environmental Modeling & Software, 17*, 721–730.

De la Rosa, D., & Van Diepen, C. (2003). Qualitative and quantitative land evaluation. In W. Verheye (Ed.), *1.5 Land Use and Land Cover, Encyclopedia of Life Support System (EOLSS-UNESCO)*. Oxford: Eolss. Available at http://www.eolss.net.

De la Rosa, D., Mayol, F., Diaz-Pereira, E., Fernandez, M., & De la Rosa, D., Jr. (2004). A land evaluation decision support system (MicroLEIS DSS) for agricultural soil protection. *Environmental Modeling & Software, 19*, 929–942. Available at http://www.microleis.com.

Derpsch, R., & Benites, J. (2003). Situation of conservation agriculture in the world. In: *Proceedings of the Second World Congress on Conservation Agriculture: Producing in harmony with nature*. Iguasu, Brasil/Rome: FAO.

Dexter, A. R., & Bird, N. R. A. (2001). Methods for predicting the optimum and the range of soil water contents for tillage based on the water retention curve. *Soil & Tillage Research, 57*, 203–212.

Dexter, A. R. (2004). Soil physical quality. Part I. Theory, effects of soil texture, density and organic matter, and effects on root growth. *Geoderma, 120*, 201–214.

Doran, J. W., & Jones, A. J. (1996). Methods for assessing soil quality. *SSSA Special Publication* 49. Madison, WI: Soil Science Society of America.

Doran, J. W., Sarrantonio, M., & Liebig, M. A. (1997). Soil health and sustainability. *Advances in Agronomy, 56*, 1–54.

EC. (2002). Towards a thematic strategy for soil protection. Communication from the EC to the European Parliament. COM 2002, 179 final. Available at http://europa.eu.int/scadplus/printversion/en/lvb/l28122.htm.

FAO. (1976). A framework for land evaluation. *Soils Bulletin, 32*. Rome: FAO.

FAO. (1978). Report on the Agro-ecological Zones Project. *World Soil Resources Report, 48*. Rome: FAO.

FAO-CSIC, (2003). The multilingual soil profile database SDBm Plus. *Land and Water Digital Media Series, 23*. Rome: FAO.

Gomez, J. A., Giraldez, J. V., Pastor, M., & Fereres, E. (1999). Effects of tillage methods on soil physical properties, infiltration and yield in an olive orchard. *Soil & Tillage Research, 52*, 167–175.

ICASA. (2006). International Consortium for Agricultural Systems Application. Available at http://www.icasa.net. (Retrieved on 20 February 2006)

Imeson, A., Arnoldussen, A., De la Rosa, D., Montanarella, L., Dorren, L., Curfs, M., Arnalds, O., & Van Asselen, S. (2006). *SCAPE: Soil conservation and protection in Europe*. The way ahead. Luxembourg: CEE-JRC.

Karlen, D. L., Mausbach, M. J., Doran, J. W., Cline, R. G., Harris, R. F., & Schuman, G. E. (1997). Soil quality: A concept, definition and framework for evaluation. *Soil Science Society of America Journal, 61*, 4–10.

Martinez-Raya, A. (2003). Evaluacion y control de la erosion hidrica en suelos agricolas en pendiente, en clima mediterraneao. In: R. Bienes & M. J. Marques (Eds.), *Perspectivas de la degradacion del suelo* (pp. 109–122). I Simposio Nacional de Erosion de Suelos.

Moreno, F., Murillo, J. M., Pelegrin, F., & Giron, I. F. (2006). Long-term impact of conservation tillage on stratification ratio of soil organic carbon and loss of total and active CaCO3. *Soil & Tillage Research, 85*, 86–93.

Moscatelli, G., & Sobral, R. (2005). *Avances en la selección de indicadores de calidad para las series de suelos representativas de la region Pampeana, Argentina*. Buenos Aires: INTA. Available at http:// www.inta.gov.ar/mjuarez.

Nortcliff, S., (2002). Standardization of soil quality attributes. *Agriculture, Ecosystems & Environment, 88*, 161–168.

Oxley, T., McIntosh, B. S., Winder, N., Mulligan, M., & Engelen, G. (2004). Integrated modelling and decision support tools: A Mediterranean example. *Environmental Modeling & Software, 19*, 999–1010.

Pachepsky, Y., & Rawls, W. J. (2004). *Development of pedotransfer functions in soil hydrology. Development in Soil Science, Vol. 30*. Amsterdam: Elsevier.

Quilchano, C., & Marañon, T., (2002). Dehydrogenase activity in Mediterranean forest soils. *Biological Fertility Soils, 35*, 102–107.

Rossiter, D. (2003). Biophysical models in land evaluation. In W. Verheye (Ed.), 1.5 Land Use and Land Cover, *Encyclopedia of Life Support System (EOLSS-UNESCO)*. Oxford: Eolss. Available at http://www.eolss.net.

Sojka, R. E., & Upchurch, D: R. (1999). Reservations regarding the soil quality concept. *Soil Science Society of America Journal, 63*, 1039–1054.

Storie, R. E. (1933). An index for rating the agricultural value of soils. *California Agricultural Experimental Station Bulletin, 556*.

Thies, J. E. (2006). Measuring and assessing soil biological properties. In: N. Uphoff, A. Ball, E. Fernandes, H. Herren, O. Husson, M. Laing, Ch. Palm, J. Pretty, P. Sanchez, N. Sanginga & J. Thies (Eds.), *Biological approaches to sustainable soil systems* (pp. 655–670). Boca Raton, FL: Taylor & Francis/CRC Press.

Thysen, I. (2000). Agriculture in the information society. *Journal of Agricultural Engineering Research, 76*, 297–303.

Uphoff, N. (Ed.) (2002). *Agro-ecological innovations*: Increasing food production with participatory development. London: Earthscan.

Uphoff, N., Ball, A., Fernandes, E., Herren, H., Husson, O., Laing, M., Palm, Ch., Pretty, J., Sanchez, P., Sanginga, N., & Thies, J. (Eds.) (2006). *Biological approaches to sustainable soil systems*. Boca Raton, FL: Taylor & Francis/CRC Press.

USDA. (1961). Land capability classification. *Agriculture handbook 210*. Washington, DC: U.S. Government Printing Office.

USDA. (2006). Soil Quality Institute. Natural resources conservation service. Available at http://soils.usda.gov/sqi/. (Retrieved on 20 February 2006.)

Van Lanen, H. A. J. (1991). Qualitative and quantitative physical land evaluation: An operational approach. Ph.D. thesis. Wageningen, The Netherlands: Wageningen Agricultural University.

Verheye, W. (1988). The status of soil mapping and land evaluation for land use planning in the European Community. In: J. M. Boussard (Ed.), *Agriculture: Socio-economic factors in land evaluation*. Luxembourg: Office for Official Publications of the EU.

Wallace, A. (1994). Soil organic matter must be restored to near original levels. *Communications in Soil Science & Plant Analysis, 25*, 29–35.

Warkentin, B. P. (1995). The changing concept of soil quality. *Journal of Soil & Water Conservation, 50*, 226–228.

Wischmeier, W. H., & Smith, D. D. (1965). Predicting rainfall erosion based from cropland east of the Rocky Mountains. *Agriculture Handbook, 282*. Washington, DC: U.S. Government Printing Office.

Wolfe, D. (2006). Approaches to monitoring soil systems. In N. Uphoff, A. Ball, E. Fernandes, H. Herren, O. Husson, M. Laing, Ch. Palm, J. Pretty, P. Sanchez, N. Sanginga & J. Thies (Eds.), *Biological approaches to sustainable soil systems* (pp. 671–681). Boca Raton, FL: Taylor & Francis/CRC Press.

Zalidis, G., Stamatiadis, S., Takavakoglou, V., Eskridge, K., & Misopolinos, N. (2002). Impacts of agricultural practices on soil and water quality in the Mediterranean region and proposed assessment methodology. *Agriculture, Ecosystems & Environment, 88*, 137–146.

Chapter 10
The Impact of Urbanization on Soils

Peter J. Marcotullio, Ademola K. Braimoh, and Takashi Onishi

Abstract Cities are important driving forces in environmental trends as a consequence of the increase in the share of the global population that reside in urban areas and the large intensity of activities of urban dwellers. As the world continues to urbanize, however, humans have lost contact with soil and the services it provides to sustain life. A review of the literature shows that the ability of urban activities to influence the physical conditions and pollution levels in soils at a distance is increasing. Cities and urban processes have had dramatic but varying impacts on soil physical and biochemical properties and pollutant loads, all of which affect the life-supporting services of soils. As developing countries continue to industrialize, soil pollutant contamination in their cities continue to increase to levels warranting immediate action. We argue for a global assessment of urban soils to identify the patterns, processes, and unique circumstances of anthropogenic impacts. There is also the need for soil protection and remediation in areas already undergoing change as a result of urban development.

Keywords Urbanization, cities, urban ecosystems, scales

10.1 Introduction

The growth and urbanization of the global human population over the last 50 years has resulted in the rapid increase in total share of global population on the one hand, and the unprecedented spatial expansion of a number of cities, on the other. While the world's megacities (i.e., those estimated to contain more than 10 million inhabitants) offer striking examples of the environmental challenges that accompany urbanization, environmental degradation has not been avoided by cities of any size (Hardoy et al., 2001). Indeed, the impact of urban activities of cities of all sizes on local, regional, and global-scale environmental trends is increasing (McGranahan et al., 2005).

Two important reasons why cities are becoming driving forces in environmental trends is the increasing share of the global population that reside in urban areas and the increased intensity of activities that these populations bring to cities. The world recently hit a level of 50% urbanization and the United Nations (UN, 2006) predicts that the share of urban population will increase to 60% over the next 30 years.

A.K. Braimoh and P.L.G. Vlek (eds.), *Land Use and Soil Resources.*   201
© Springer Science+Business Media B.V. 2008

Accompanying this shift in residence is the shift in location of productive and consumptive activities, waste products emanating from these and associated risks. International organizations, such as the World Bank, consider cities to be the engines of growth, particularly in the era of global flows of trade, finance, investments, people, and information. Common wisdom suggests that cities are and will continue to be the places where most human activities and their impacts are generated.

The roles played by cities in the generation of environmental impacts, including the perturbation of global biosphere and geochemical cycles are therefore drawing increasing attention. One area of particular importance is the impact of urban growth on environmental resources (air, water, soils, and biodiversity). Of these, soils have received increasing attention. As late as the mid-1980s, Spirn (1984) pointed out that one of the most misunderstood and least researched and documented aspects of the urban environment was soil. Certainly, studies of the physical characteristics of urban soils were undertaken prior to the 1980s. Some of the first systematic studies of soils in large cities began during the late 1970s (Craul, 1985). Studies of the impacts of urbanization on regional landscapes began even much earlier. The impact of urbanization on agricultural land became an academic concern during the 1950s (Bogue, 1956). Furthermore, a large number of studies of trace element concentrations and fluxes had been performed on urban street dust and urban soils, starting in the 1960s (Purves & Mackenzie, 1969; Purves, 1972). Nevertheless, since the 1980s, studies of urban soils have broadened and multiplied. By the late 1980s, Gilbert (1989) was able to put together a comprehensive study of the impact of cities on various environmental resources, and included an entire chapter on soils. By the early 1990s, entire volumes were being devoted to urban soils (see, e.g., Bullock & Gregory, 1991; Craul, 1992). Thereafter the literature on urbanization and its impact on soils expanded significantly. This chapter focuses on the literature related to changes in soils and urbanization. The idea is to take stock of the knowledge generated and summarize some of the consensus.

We have organized the review in the following sections. Section 10.2 presents our way of organizing the impact of urbanization-related changes on soils. This includes an introduction to the current scale of urbanization and the need for a multiscale perspective on how urban activities affect soils. Section 10.3 reviews the impacts of urbanization on the physical characteristics of soils. Section 10.4 reviews the literature on soil pollution. Finally Section 10.5 summarizes and concludes. The review is a first step in attempting to draw together an increasingly growing literature on the impact of urbanization on soils.

10.2   Framework for Understanding the Impact of Urbanization on Soils

This section briefly outlines the background and framework for the review. In the first part (Section 10.2.1), we review postwar urban population growth and the growth in the general size of cities as well as the regional distribution of both urban populations and the number of cities. In Section 10.2.2 we describe the potential impacts of urbanization and of city growth on soils. In Section 10.2.3 we present the framework for exploring the impacts of these dramatic changes.

10.2.1   The Scale of Contemporary Urbanization

Despite the fact that cities emerged some 8,000 years ago, it was only very recently that most of the world's population have resided in dense settlements. In 1800, for example, only 2% of the world's population resided in urban areas (Torrey, 2004). No society during that time could be described as predominately urbanized. By 1900, the global share grew to approximately 14% of the world's population (Douglas, 1994) and only one nation, the UK, was considered an urban society. Previous to the UK, cities in the Netherlands grew in importance but the country remained less than 40% urbanized (Berry, 1990). During the late nineteenth and early twentieth centuries, much of the developed world experienced rapid urbanization. The USA, for example, became 50% urbanized during the 1920s. It was after 1950 that the bulk of the world's population began a rapid shift from rural to urban settlements.

The postwar urban migration has been massive, intensive, and largely centered in developing countries. As Table 10.1 demonstrates, from 1950 to 2000 more than 2 billion people have moved into the world's cities, bringing the proportion of urban population up from about 30% to more than 47%. Approximately 78% of this change occurred in less developed countries. National urbanization patterns typically follow an attenuated "s-shaped" curve (Davis, 1965), meaning that once the process begins it rapidly accelerates and then slows down at approximately 60–70% urbanization levels. The urbanization curve for the world demonstrates that global urbanization is in the middle of the rapid growth period. With approximately half the world's population currently living in dense human settlements, the next 30 years will bring greater changes than previously experienced. For example, approximately 4.9 billion of the total 8.1 billion inhabitants of the Earth will end up in cities by 2030, an additional 2 billion people (representing an increase of approximately 72%) from those of today. This will increase the world's urbanization level to approximately 60%.

Postwar urbanization is different from previous trends. Over the last 50 years global urbanization has occurred at some of the fastest rates in history. In the 1950s, the world annual 5-year average increase was approximately 1.2% annually, but this figure masks the high speed of change in some countries. For example, some countries in Africa were reaching urbanization-level increases of over 10% (including Botswana, Lesotho, and Swaziland),[1] while those in Asia

---

[1] For many countries in Africa, census data are not available for several decades and therefore these increases are subject to debate (see, e.g., Satterthwaite 2005). These estimates for changes in urbanization levels are from UN (1999) data.

Table 10.1 Global Urbanization, 1950–2000 (thousands). (Adapted from McGranahan et al., 2005)

	1950	2000	Absolute change 1950–2000	Percent change 1950–2000
World Population	2,521,495	6,055,049	3,533,554	140.1
More developed regions	812,687	1,187,980	375,293	46.2
Less developed regions	1,708,808	4,867,069	3,158,261	184.8
Least developed countries	196,764	644,677	447,913	227.6
World Urban Population	749,934	2,845,049	2,095,115	279.4
More developed regions	446,138	902,993	456,855	102.4
Less developed regions	303,795	1,942,056	1,638,261	539.3
Least developed countries	14,002	167,421	153,419	1,095.7
Percent Urban	29.7	47.0	17	58.0
More developed regions	54.9	76.0	21	38.5
Less developed regions	17.8	39.9	22	124.4
Least developed countries	32.5	47.5	15	46.0
Urban Growth Rates	3.01	2.03	—	—
More developed regions	2.32	0.50	—	—
Less developed regions	3.97	2.70	—	—
Least developed countries	4.56	4.49	—	—
Urbanization Rates	1.23	0.83	—	—
More developed regions	1.11	0.30	—	—
Less developed regions	1.93	1.26	—	—
Least developed countries	2.66	2.13	—	—

exceeded 2–3% annually (including North Korea, Afghanistan, Iran, Nepal, and others). The growth of the share of urbanized population in developing and least developed countries demonstrates a faster urban transition than experienced by the developed world.

Another important difference between contemporary urbanization and historical patterns is the absolute size of populations entering cities. As Table 10.1 demonstrates, from 1950 to 2000 the developing world urban population expanded by 1.6 billion. In 1950, the total world population stood at only 2.5 billion.

A third feature of current urbanization is the rise of large urban agglomerations. The total number of large cities (i.e., those greater than 1 million) climbed from 85 in 1950 to 405 in 2000. The largest percent increase among the different sizes of cities was the "megacities" category (i.e., those equal to or greater than 10 million inhabitants). According to the UN (1999), there was one megacity in 1950, but this number increased to 19 by 2000. Of the 19 megacities in the contemporary world, 11 (i.e., 58%) are located in Asia, and most in developing Asia.[2]

---

[2] Historically large cities have been associated with large economies. For example, before 1820 the currently developing world accounted for 80% of world GDP and with that the location of the majority of large cities (Chandler 1987; Maddison 2001). It was only during the industrial revolution and the growth of western economies when large cities blossomed in these parts of the world. As the developing economies grow economically, they cities are retaking their place amongst the global hierarchy.

The number of large cities will continue to increase in the medium term. The UN predicts that by 2015, the number of megacities will increase to 23. In the next 15 years, however, it is the middle-size large cities (i.e., those from 5 to 10 million inhabitants) that will gain the largest numbers and increase in relative size. It is interesting to note that while the proportion of the urban population is expected to dramatically rise, the proportion of those living in megacities will not. Large cities are growing more slowly than others. The UN projects that of the total urban population, the megacities will continue to maintain a share of approximately 12% in 2030, a fraction slightly higher than that of 2000. Most of the urban population (approximately 60%) will continue to live in cities with less than 1 million inhabitants.

Urbanization, as land cover, in the form of built-up or paved-over areas, occupies less than 3% of the earth's land surface (McGranahan et al., 2005).[3] Urban land uses have also been concentrated in specific ecosystems. Table 10.2 shows the proportion of urban population, densities, and land uses within various ecosystems around the world (see McGranahan et al., 2005). The table demonstrates the uneven distribution of cities by ecosystem (as the ecosystem types are not mutually exclusive, the data present only general conclusions). The largest share of the urban population (68%) lives in cultivated ecosystems.[4] This is not surprising as cultivated systems accounted for 24% of the Earth's terrestrial surface. The next largest share of the global urban population (34%) lives in drylands. The third highest proportion of the urban population lives in coastal zone and inland waters ecosystems (together they account for 54% of the total urban population). Of particular note is the high densities found in coastal systems, along with the large proportion of the system given over to urban use (10.8%).

The urban populations within each of these ecosystems can be grouped together by different-size cities (Table 10.3). Coastal zone urban areas have the highest percentage of the world's large cities. Over 65% of the urban population residing in cities within coastal ecosystem lives in cities larger than 1 million persons. In cities located in cultivated and dryland ecosystems, most people live in cities of between 5,000 and 1 million inhabitants (53% and 57%, respectively). Given the distribution of the world's urban population between developed and developing, and between different types of ecosystems, we would expect differential impacts on soils.

10.2.2   *The Impact of Urbanization on Soils*

Soil is the productive layer of the earth's surface. It evolves with the biotic community based upon climatic conditions, local biota, topography, parent material, and

---

[3] Interestingly, Doxiadis and Papaioannou (1974) predicted that the world's urban land cover would reach 2,770,408,000,000,000 m^2 (343 times the size of megalopolis) and approximately 2% of the Earth's land surface.

[4] Cultivated ecosystems are defined by those areas where at least 30% are under agricultural production.

Table 10.2 Global population estimates and densities in rural and urban regions. (Adapted from McGranahan et al., 2005)

Ecosystem	Population estimates (million)				Land areas (× 1,000 km²)				Population density (persons km⁻²)			Proportions	
	Total	Urban	Rural	% Urban	Total	Urban	Rural	% Urban	Overall	Urban	Rural	Land use	Population
Coastal zone	1,147	744	403	64.9	6,538	665	5,873	10.2	175	1119	69	18.1	26.3
Cultivated	4,223	1,914	2,309	45.3	35,476	2,413	33,063	6.8	119	793	70	65.7	67.7
Dryland	2,149	963	1,185	44.8	59,990	1,286	58,704	2.1	36	749	20	35.0	34.1
Forest	1,126	401	725	35.6	42,093	839	41,253	2.0	27	478	18	22.8	14.2
Inland Water	1,505	780	726	51.8	29,439	944	28,496	3.2	51	826	25	25.7	27.6
Mountain	1,154	349	805	30.3	32,084	549	31,535	1.7	36	636	26	14.9	12.3
Overall	6,052	2,828	3,224	46.7	130,670	3,673	126,996	2.8	46	770	25	100	100

Notes: Population estimates have been rounded to the nearest million. Population numbers for each ecosystem will not add to total as numbers have been rounded and systems are not mutually exclusive.

Table 10.3 Population and percent share in urban areas of various population sizes within select Millennium Ecosystem Assessment systems. (Adapted from McGranahan et al., 2005)

Area	Urban population ('000s)	Urban settlement size								
		5,000–20,000	20,000–50,000	50,000–100,000	100000–500000	500,000–1,000,000	1,000,000–5,000,000	5,000,000–10,000,000	>10,000,000	Larger than 1 million
Coastal zone										
Population	744,000	13,000	28,000	33,000	112,000	69,000	196,000	119,000	175,000	
Percent		1.7	3.7	4.4	15.0	9.2	26.4	16.0	23.5	65.9
Cultivated										
Population	1,914,000	75,000	175,000	166,000	411,000	183,000	484,000	172,000	249,000	
Percent		3.9	9.1	8.6	21.5	9.5	25.3	9.0	13.0	47
Dryland										
Population	963,000	39,000	84,000	88,000	224,000	111,000	260,000	71,000	85,000	
Percent		4.1	8.7	9.2	23.3	11.5	27.0	7.4	8.9	43
Forest										
Population	401,000	22,000	43,000	37,000	83,000	41,000	98,000	26,000	52,000	
Percent		5.5	10.7	9.2	20.8	10.1	24.3	6.4	12.9	44
Inland Water										
Population	780,000	24,000	49,000	48,000	151,000	81,000	193,000	79,000	154,000	
Percent		3.1	6.2	6.1	19.4	10.4	24.8	10.2	19.8	55
Mountain										
Population	349,000	21,000	47,000	35,000	77,000	34,000	85,000	25,000	24,000	
Percent		6.1	13.3	10.1	22.1	9.7	24.4	7.2	7.0	39

Notes: Urban population figures have been rounded to the nearest million, therefore total population does not equal to the sum of populations in all settlement sizes. Percent columns do not sum to 100. Island systems not included.

time. Soil is a natural system with structure and specialized features (Rozanov et al., 1990). Because soils are productive, they are crucial to life on earth. This is well understood by farmers and those that work closely with soil. As the world is increasingly urbanized, however, people have lost contact with soil and with that they have lost an appreciation of how important soils are to their everyday lives. This tendency is ironically counterpoised against increased dependency urban populations have on soils for food, for waste recycling, and as carbon sinks among other functions.

The question arises as to how to assess the impacts of living in cities on soils. In order to understand these impacts, we first need to understand the basic role of soils in our ecosystems. Brady and Weil (2002) describe five functions of soils including: supporting the growth of higher plants, controlling water within the hydrologic system, mediating nutrient cycles, providing habitat for living organisms, and serving as an engineering medium. From these functions it is then possible to identify how urbanization might impact soils.[5] First, soils have a physical structure, temperature, and pH that help to provide support for higher plants. Second, soils control the cycling of essential nutrients and micronutrients, water movement, and they also act as a sink for wastes of all sorts. Finally, soils support biodiversity. These three characteristics therefore, can serve as a basis for an assessment of changes in soil; namely an impact assessment should identify the changes in physical structure, soil-related biochemistry, and/or soil biodiversity.

Identifying the focus of impact is not enough to undertake the review, however. Soils cover most terrestrial ecosystems, and we therefore need to target the locations impacted by urban activities. Many previous studies examined the soils within urban borders. These studies restricted their perspective of the impacts of urbanization understandably, but artificially to the local scale. Cities and urbanization processes, however, affect soils both directly and indirectly outside city boundaries.

Increasingly, urban activities are seen to impact larger scales. Scale, in this sense translates into geographic extent or temporal duration of the process. Scale is often referred to as local, regional, or global. Local scale refers to particular phenomena, processes, or activities that are most evident at household, neighborhood, or community level. For example, in environmental terms, indoor air pollution and its impact is a local phenomenon because it is best observed and measured at that level. While indoor air pollution statistics may be aggregated at the global level, understanding the processes that created it, its uneven distribution, and its impacts on human health are less obvious at regional or global scale. Alternatively, the environmental impact of carbon dioxide emissions is often referred to as a global-scale issue. This is because total global amounts of this gas have systemic impacts on the Earth's climate. Observing and measuring changes in this emission and its impacts are best performed at this scale, although there are also important regional and local impacts.

---

[5] These were similar but not exactly alike to the general categories of Oldeman, et al.'s (1991) study including chemical change, erosion, salinization, compaction, and pollution, which are used to evaluate the present status of the world's soil resources as affected by human impact.

Urban development and economic growth have been associated with the changing scale of impacts (McGranahan et al., 2001). That is, cities' transition from one set of environmental challenges (including impacts) to another as they grow in wealth. This theory, entitled, urban transition theory, is powerful in its ability to disaggregate the types and dominance of environmental conditions associated with cities at different levels of income and how these conditions change in terms of their type, geographical, and temporal extent. The theory identifies three important scales of analysis—local, citywide (or regional), and global. In this review, we consider these three different scales of urbanization's impact on soils.

10.2.3   Assessment of Soil Impact by Type and Scale

Scientific studies of soils vary in terms of both subject and methodologies. Moreover, not all knowledge of the impact of urbanization on soils can be gleaned from studies performed specifically on urban soils. At least two challenges emerge for a scaled analysis of urbanization's impacts on soils. First, a global assessment of urban soils is not available. Researchers attempting to generalize findings to a global level, therefore, must either overlook differences between soil environments within and between development contexts or identify nonsoil indicators that reflect impacts (such as percent of impervious surfaces). Second, larger-scale perspectives of urbanization's impact on the environment present challenges in terms of identifying linkages between activities and environmental outcomes. Nevertheless, studies exist that attempt to do both. Below we describe the types of studies reviewed at each scale.

10.2.3.1   Studies Identifying the Impacts of Urbanization at the Local Scale

Urbanized land is not completely paved nor is it uniform. In the USA, larger residential properties (>4,000m²) have less impervious coverage (20%) while smaller residential properties (500m²) can have up to 65% sealed surfaces. Industrial, commercial, and shopping centers (strip-type commercial development) are mostly impervious (75%, 85%, and 95% paved surfaces, respectively) (US Natural Resources Conservation Service 1986). Globally, Grubler (1994) suggests that built-up areas (land for building and infrastructure) within cities typically range between 200 and 250 m² per capita. Even in dense countries, such as Japan and the Netherlands, he estimates that built-up land is approximately 11% and 4% of total land area, respectively. Areas within cities with open soils have been studied by Rowntree (1984), who found that in the USA, four cities demonstrated tree-canopy cover between 24% and 37% of total area. Suburbs accounted for 46% of total urban areas and here on average 14% were vacant lands. Sanders and Stevens (1984) found that within Dayton, Ohio, approximately 50% of land use had canopy cover, and 2 km² were open soil. Urban tree cover in the USA ranges from 0.4% in

Lancaster, California, to 55% in Baton Rouge, Louisiana (Nowak et al., 1996). Paris, one of Europe's greenest cities, has approximately 450 parks and gardens covering more than 3,000 ha accounting for almost 30% of the city's surface (Ackerman, 2006). Within Vienna only 33% of land is for offices and public buildings and infrastructure (Grubler, 1994). Sealed soils in all urban Germany include 52% of the built-up areas (European Environment Agency, 2006). Jim (2003) estimated that urban development in Hong Kong is concentrated on 20% of the land, leaving the rest as rural hinterland. These differences represent possible differential impacts of urbanization processes upon soils within cities. Therefore the most reliable studies of the local impact of urbanization on soils have been performed within individual parts of cities (roadsides, parks, industrial areas, etc.). Caution must be taken when attempting to generalize these findings for entire city areas.

The study of soils within cities has followed two major perspectives. The first perspective is temporal, involving direct observations and assessment of changes in soil quality over time or at a specific moment (see, e.g., Francek, 1992; Gbadegesin & Olabode, 2000; Manta et al., 2002; Ordonez et al., 2003). The second perspective is spatial, including urban–rural gradient analyses (see, e.g., McDonnell et al., 1997). Typically soil samples are collected along transects defined from the center of an urban area to the periphery, analyzed for biological, chemical, and physical properties, and correlated with the degree of urbanization. As urbanization occurs over time and results in a complex set of effects, gradients, by substituting space for time, allow for a view of what happens to soils during city growth. Those samples from rural areas are expected to have the least impact from urbanization while samples within the central city experience the greatest impact.

Two sets of limits of generalizations from these perspectives exist. The first set concerns the studies within individual cities or parts of cities. Here, direct analysis of soils within parts of different cities often uses different methods of data collection and laboratory analyses, making cross-city comparisons difficult. Furthermore, soils within and between cities may vary, again making comparisons difficult. The second set of limitations concern the urban–rural gradient analyses. In these cases, other factors, such as prevailing wind direction may play an important role in the quality of soils. A single transect from the center of a city to rural areas may miss the impact of winds carrying pollutants from the center, for example. Also, as with the caveat to direct observation studies, soils along the urban–rural gradient may intrinsically differ significantly.[6] This would make comparisons between urban and

---

[6] Niemela et al. (2002) argue differently. They suggest that the gradient occurs all over the world and provides a useful framework for comparative work on a global scale. Gradients, in their argument represent anthropogenic patterns and processes. While this may be true, anthropocentric impacts vary from city to city as do natural factors (climate, hydrology, soils, etc.). Both combine to create the state of the urban environment. As a result comparing, for example, the development (both socioeconomically and biophysically) of Baltimore, MD, with that of Phoenix, AZ, becomes difficult. The Niemela et al. (2002) study found weak and varied responses of biodiversity to urban–rural gradients across cities and indeed concluded that local factors and their interactions were of primary importance to community composition.

rural soils at best difficult, or at worst meaningless. Nevertheless, direct observation and urban–rural gradient studies provide data on the state of the environment within cities as well as on the changes to soils with growing urbanization.

10.2.3.2 Studies Identifying the Impacts of Urbanization on Soils at the Citywide and Regional Scale

Studies on the citywide impacts of urbanization on soils also fall into two categories. The first type of study includes material and energy flow accounting studies. Material flows accounting (MFA):

> tracks the amounts of materials—as classes or individual substances—that enter the economy, accumulate in capital stock such as housing and automobiles, or exit to the environment as waste. MFA documents the commercial life cycle of materials that become part of the industrial economy, from extraction, processing, and manufacturing to use, reuse, recycling or disposal. (Wernick & Irwin, 2005, p. 1)

MFA analyses not only provide data and indicators of interest to this study, but also provide a way to understand urbanization's impacts on soils.

These studies can be used to estimate the magnitude of different materials that remain in cities (inflows minus outflows) (Decker et al., 2000). In this context, MFA studies have an added advantage of identifying the context (i.e., socioeconomic activities) under which change occurs. For example, material and energy flows into building construction demonstrate the enormous amount and types of materials that are needed to build these structures. They strongly suggest the pathways for the concentrations of these materials both pre- and postconstruction, leading to an understanding of how and why soils within cities develop as they do. Impacts occur through socioeconomic activities of transforming and transferring goods, energy, and services (Douglas, 1981). Decker et al. (2000) provided a review of studies that examined cities in this light. They classified material flows into inputs (stored, transformed, and passive) and outputs (atmospheric and aquatic and marine). Stored inputs become part of the built environment, such as stone, wood, and metals, or are used until they become waste products "stored" in landfills through both active and passive processes. Transformed materials include water, food, and fuel, which undergo change from one form to another before they are exported from the urban system as output wastes. Passive inputs include precipitation, water surface flows, gas and airborne particle movements, energy from solar radiation, and movements of plants and animals. All these processes impact the quality of soil both within and outside the city boundaries. While national material flow studies began during the 1960s when societal concerns focused on pollution of environmental resources by humans, recent studies have been performed on individual cities including New York (Wolman, 1965), Hong Kong (Newcombe et al., 1978), and Singapore (Schulz, 2005).

The second set of studies that can be used to assess the impact of urbanization on soils include those that examine urban-scale land-use and land-cover change. Changes in the area of urban land per se, currently do not appear to be central to land-cover change. Concluding, however, that the small area of land actually covered by structures of settlements converts into a small impact on the environment, however, would be a mistake. Douglas (1994), for example, concluded that the importance of human settlements in environmental change is related less to the area covered with buildings and other structures, than to the role of settlements and settlement-related activities in promoting land-cover change generally, and the varying rates of conversion of land to urban and rural settlement uses. Therefore, in understanding the role of cities in environmental change, we need to look further than the amount of land covered by cities, although in some cases this can be significant as well.

Moreover, cities grow outwardly at rapid rates. In 1777, when Dr. Johnson remarked that the man who was tired of London was tired of life, open country was 3 km from the center of town and the town probably housed 750,000 inhabitants. Twentieth-century Greater London, at its height, held over eight-and-a-half million people and covered about 1,600 km². Its built-up area in parts was up to 48 km across. The expansion of New York and Paris are presented in Fig. 10.1. The data

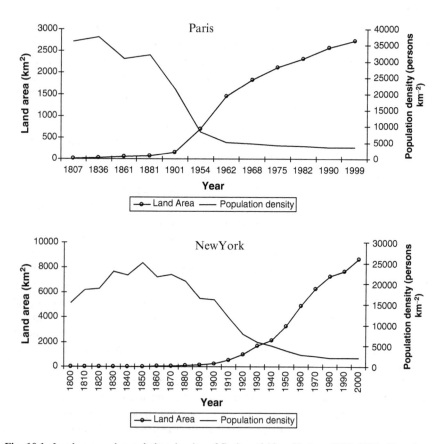

Fig. 10.1 Land area and population density of Paris and New York ca.1800–2000. (Based on http://www.demographia.com/db-parisua.htm. From 1950 to 2000, estimates for New York based on US Census Bureau)

demonstrate the large increase in size of these urban areas over time and the decrease in population density after the turn of the nineteenth century.

Contemporary developing world cities are both different from those of the developing world and also grow at different rates. Studies comparing urban development demonstrate different patterns between developing and developed countries (Huang et al., 2006). Latin American urban form studies demonstrate the compactness and density of cities related to the influence of European colonial planning. Studies of Asian cities show that they are also compact and dense, with a dominant large core area, although there is wide variation. In Korea, for example, controls on exurban land use have had similar effect as those in Japan (Yokohari et al., 2000). Moreover, in rapidly developing Chinese cities, increases in urban land use have been significant over the last few decades. For example, in over a 20-year period, from 1978 to 2002, Chengdu almost tripled in size from $81\,km^2$ to nearly $220\,km^2$ (Schneider et al., 2003).[7]

Underpinning these analyses is the perspective that cities are different and new types of ecosystems. They have, for example, particular climatic characteristics and different topographies, created by human-made structures, which in turn have diverse ecological effects (Gill & Bonnett, 1973). This viewpoint also considers cities to be made up of a patchwork of ecologically heterogeneous systems. Landscape ecology addresses hierarchical patch dynamics associated with fundamental units of landscapes and their interaction at different scales. A fundamental assumption in the field is that spatial patterns have influence over the flows of material, energy, and information through ecosystems and that these flows and the processes that create them impact spatial patterns (Wu & Hobbs, 2002). Understanding landscape structures and how they impact various environmental media, therefore is one of the "Top 10" landscape ecology research priorities (Wu & Hobbs, 2002).

Land-use and land-cover change as well as landscape ecology studies require the mapping of landscapes, identifying and quantifying flows of materials, and associating these with soils. Geographic information systems (GIS) and remote sensing are important tools for identifying patches and for linking spatial data with nonspatial data, bridging scales, and identifying areas of change and degradation (Li et al., 2004). Specifically for soils, Braimoh et al. (2005), use GIS to identify associations among soil variables that were impacted by different land uses. These types of studies demonstrate the importance of including mapping analyses in urban soil research at this scale.

10.2.3.3 Studies Identifying the Impacts of Urbanization at the Global Scale

Regional- and-global scale analyses do not directly relate the impacts of urbanization to environmental conditions, but rather estimate impacts through modeling. There are at least two types of modeling used for these types of studies. The first include

---

[7] The authors of this report state that smaller cities around Chengdu increased their land sizes at even faster rates.

scientific studies designed to identify how much "nature" humans need in order to survive in our current manner. These can be further subdivided into, at least, two types: the measures of human appropriation of net primary production, and measures of a population's Ecological Footprint (EF) (Wackernagel & Rees, 1996).[8]

The EF is "the area of biologically productive land and water required to produce the resources consumed and to assimilate the wastes generated by that population using prevailing technology" (Wackernagel & Rees, 1996). The concept turns the need to identify "carrying capacity" of a human system on its ear by providing a method for estimating the biologically productive area necessary to support current consumption and waste patterns. One can think of the EF as being the sum up of specific real locations, and the "distant elsewheres" that support us (Rees, 2002). Hence, when we address the impacts of urbanization, we must look to wider areas than those within the city's boundaries (Rees, 1992; Wackernagel & Rees, 1996). In order to demonstrate this outcome, the EF uses spatial units that are mutually exclusive from all others. Thus, the concept suggests that, "people are competing for ecological space". For most industrial regions, the ecological footprint is composed of several different land uses, resources, and ecosystem services, including crop and pasture land; built-up land; forest; fish and carbon assimilation. The footprints of these regions exceed the capacity available locally. This leads to that population's appropriation of ecosystem goods and services from that of the global system. In fact, recent work indicates that the total global ecological deficit is rising and that it now takes the global biosphere 1.2 years to regenerate what humans use in 1 year (Wackernagel et al., 2002).

Studies of the appropriation of nature have also demonstrated the large resource consumption (and hence impact) of cities. Economically powerful urban consumers arguably have impacts on environments distant from cities. Savage (2006), for example, argues that the growing number of wealthy people in Asia that can pay high prices for goods, create incentives to capture some marine species (even those endangered) with techniques that destroy wider systems (i.e., cyanide bombs within coral ecosystems).

Other studies have attempted to empirically estimate the impact of cities on wider ecosystems. Inhabitants within urban areas of the Baltic Sea drainage, for example, depend on forest, agriculture, wetland, lake, and marine systems that constitute an area about 1,000 times larger than that of the urban area proper (Folke et al., 1997). In terms of the impacts on soils, for example, it is possible to associate the impact of urban consumption levels on tropical soils to deforestation and erosion. A recent study, of tropical hardwoods consumption in the state of Sao Paulo demonstrates that one in every five trees in the Amazon makes its way to the city. Braimoh (2004) indicates that woodcutting for charcoal production and the associated impact on soils in semirural Northern Ghana is largely driven by the energy demand in Tamale city, the northern regional capital.

The other side of consumption is waste production. There are also models that identify the impact of urban emissions on distant soils. For example, Chameides

---

[8] For a comparison of these two types see Haberl et al. (2004).

et al. (1994, 1999) identified the impact of ozone emissions from "continental-scale metro-agro-plexes" (CSMAPs) on soils and therefore agricultural production.

The second set of studies that identifies the global impacts of cities on soils includes land-use and land-cover change at the global scales. Several recent global mapping projects provide basic global spatial data for these studies. They help to identify the location of cities within specific ecosystems, for example, agricultural and coastal zones.

10.3   Urbanization's Impacts on Physical Properties of Soils

Human activity within cities changes the physical character of soils. Material inputs into cities have exceeded output for several years. This has meant several things, including the general rise of ground level in older districts as well as the infill of waterways, streams, and valleys in other parts of the city. With the constant changes and additions made to the urban ecosystem, it is no surprise that soils located within urban areas are unique.

The physical characteristics of soils include texture (particle size), bulk density, structure (shape of peds), plasticity, permeability, porosity, temperature, and moisture content. Craul (1985) was perhaps the first to systematically study the changes in soils within cities. Much work thereafter has been based upon his initial observations. This part of the review starts with some of the observations made by Craul (1985, 1992) and his successors. Work to date suggests that the distinction between urban soils and those more "natural" soils is so great that soil scientists have classified them differently.

At the regional level, urbanization has been associated with the advance of impervious surfaces that have sealed soil from the atmosphere and taken significant amounts of land from agricultural production. These changes have led to shifts in the hydrological cycle, as infiltration capacity is reduced, changing river and stream flow and sedimentation levels. Moreover, development in peri-urban areas has been associated with massive gully erosion. On the other hand, in developing countries access to markets (meaning the growth of towns and cities) can have an effect on soil quality, as farmers then can buy fertilizers and soil supplements. At the national and global level, urbanization and the accompanying changes in consumption patterns play a role in nutrient depletion of rural soils and accelerated erosion due to the expansion of urban areas. On the other hand, in some cases, rural–urban migration (particularly in developing countries) leads to land abandonment and forest regeneration.

10.3.1   Local-Scale Impacts of Urbanization on Soils

Craul (1985, 1992) and Gilbert (1989) summarized differences in the physical properties of urban and nonurban soils. They suggested that urban soils were

Table 10.4 General differences between urban and more natural soils (Adapted from Craul (1992))

Characteristics of urban soils in comparison with natural soils	Causes	Resulting problems
Harsh boundaries between soil layers	Artificial origins produce layering of different materials	Lack of continuity for tooting plants and burrowing soil animals
Compaction	Trampling and pressure from vehicles, etc.	Reduced water passage and lack of air spaces. Plants produce shallow roots
Low water drainage	Diversion of runoff to drains, and interruption of natural flow through soils	Reduced water availability for plants
Crusting and water repellency	Compaction, chemical dispersion, and creation of waxy soil surface	Barriers to gaseous and water exchange between soil and atmosphere
High pH	Effects of de-icing salts and water running over calcareous building materials (e.g., concrete)	Problematic if highly alkaline, because some nutrients (e.g., phosphates) are immobilized
High soil temperatures and moisture regimes	Higher ambient air temperatures and little buffering effect of vegetation	Reduce moisture in upper layers for plants growth

characterized by reduced soil structure, compaction, surface crusting, restricted aeration and drainage, and modified pH and temperature regimes (Green & Oleksyszyn, 2002). Table 10.4 summarizes each of these effects, their causes, and resulting problems. A detailed discussion is provided below.

10.3.1.1 Heterogeneous Soil Structure

One of the end products of the natural process of soil formation is the development of soil structure; the qualitative description of peds, pores, and crack geometry; and spacing within soils (Bullock and Gregory, 1991). The process of aggregating primary particles into peds results in a net increase in bulk soil volume and the associated lightening of the soil (a decrease in bulk density) through the creation of a greater proportion of macro-pore space.

Contrary to natural and slightly modified soils, urban soils often display a succession of layers that are not always parallel to the soil surface. Urban soil profile typically displays marked variability comprising broken horizons in the vertical plane. In many profiles, horizons gradually grade from one to the next and differentiation between horizons is sometimes difficult. Urban soils can also change abruptly at one or more levels within the profile. Craul (1992) demonstrates several examples of urban soil profiles exhibiting such *lithologic discontinuities*. Spatial variability can be just as complex as vertical variability, drastically impacting the rooting environment (Gilbert, 1989). This is due in large part to the rapid transformations of materials from one form to another within the urban environment at much faster rates than those experienced in natural soil-forming processes. Urban activities employ modern machinery such as tractors and bulldozers, which within days can completely change an entire landscape. The digging required for new buildings, leveling and preparing land for new uses, disposal of rubbish and debris, or application of topsoil for landscaping, hinders the development of a relationship between the urban soil and its parent material. On some derelict sites within older parts of cities, Tower Hamlets in London for example, there is little natural soil remaining and the surface of the ground is covered with demolition debris from buildings previously occupying the site (Bullock and Gregory, 1991).

Under these conditions, urban climate and local fauna also have little to do with soil formation. In order for scientists to reconstruct the history of the soils in cities, efforts require collaboration amongst soil scientists, historians, and archaeologists (De Kimpe & Morel, 2000). Studies in Bangkok, a postwar, rapidly expanding city suggest that soils laying along a north–south corridor running from suburbs to industrial zones (though the city center) can be between 2 and more than 30 years old (Wilcke et al., 1998).

In other sites, the urbanization process increases the spatial heterogeneity of the soil-scape by creating recently altered or modified soils in close proximity to older, naturally occurring soil bodies (Craul, 1992). Short et al. (1986a) noted that while soils in the Mall in Washington, DC, were fill deposited in repeated applications, B horizons were recognized in 26% of samples of a Mall in Washington, DC, while A horizons were found buried in 42% of the profiles. Moreover, they found mean organic matter content increases from the surface to the sixth horizon hypothetically related to the presence of buried A horizons and of fill material containing variable amounts of organic matter. High organic matter in subsoils were also found in Stuttgart, Germany (Lorenz & Kandeler, 2005). Microbial biomass contained 0.12 g C kg^{-1}, and 0.05 g N kg^{-1} at 1.7–1.9 m depth, which in some samples was higher than in layers closer to the soil surface.

Scheuss et al. (1998) found young soils in urban areas of northern Germany, which demonstrated strong spatial heterogeneity. The soils often contained large amounts of coarse material and macro-pores ($>60 \mu m$ diameter). Most of the study area included soils that developed from a mixture of garbage and rubble with sandy materials. Jim (1998a), in 80 roadside soil pits observed sharp boundaries between layers, filled by decomposed granite contaminated by construction rubble. Gilbert (1989) found brick rubble sites in inner Liverpool and Belfast which contain texture as good as topsoil. There is a large proportion of the sand-sized fraction (0.02–2.0 mm diameter). Rubble soils are also dominated by the coarser fraction of substrates (gravel, cobbles, and stones). In this case, the profiles revealed very little organic matter.

The heterogeneity of urban soils can be also very large within any one layer (De Kimpe and Moral, 2000). In Ibadan, Nigeria, the mean level of the depth of disturbance is about 15 times that of the rural soils (Gbadegesin & Olabode, 2000). Given these characteristics, understanding the development of urban soils requires reconstructing the history of building and land-use patterns within the area.

Finally, cities located on coastal areas have partially developed on fill. Approximately 81% of Washington, DC, area is disturbed land, 14% of which is classified as fill (Craul, 1992). Coastal landfill can be made from previous waste disposal sites, construction-material waste, soils taken from development of other areas in the city, or fresh soils from outside the city. Many coastal cities in the rapidly developing world engage in harbor and coastal in-filling (Lo & Marcotullio, 2000)

10.3.1.2 Compaction

Compaction is the process of reduction of the specific volume (or porosity) of a soil. Soil compaction inhibits drainage, aeration, and root growth and thus has received attention in the agricultural literature (Bullock & Gregory, 1991). Compacted soil often behaves like impervious surfaces, concrete, or asphalt. Good agricultural soil has about 50% pore space. When a soil is compacted so that it has less than 25% pore space, typical of urban soils, it becomes a poor medium for supporting plant growth.

Forces acting in opposition to compaction are frost heave and the effects of soil fauna. Compaction can also be prevented through management (e.g., plowing). In urban areas, however, typical conditions tend to destroy structure and increase bulk density. Craul (1992) identified six conditions within urban areas that promote compaction including:

1. Urban soils demonstrate, at least, partial destruction of their structure and horizon arrangement within the profile, enhancing compaction. Conversely, the absence of periodic tillage helps facilitate compaction.
2. Urban soils have low organic matter, which is an aggregating, structure-forming agent.
3. Urban soils with low organic matter have limited soil organism populations which promote soil structure and increase soil porosity.
4. Elevated urban temperatures reduce the frequency of complete freeze–thaw cycles, thus further preventing soil structure formation.
5. Urban soil surfaces are subjected to various physical activities over a range of moisture conditions, destroying vegetative cover and leaving the soil surface bare and unprotected from compacting forces, while eliminating the binding effect on soil particles by plant root systems.
6. Urban compacted soils have different wetting and drying cycles from uncompacted soils. Compaction is further exacerbated when wet through surface traffic by foot or machine.

Bulk density is the measure used to determine the degree of compaction and indicates how closely the soil particles are packed together. Bulk density is often expressed as porosity (the volume of voids per unit volume of soil) in terms of mass per unit volume; grams per cubic centimeter ($g\ cm^{-3}$). Well-aggregated soils, rich in organic matter, have bulk density values less than $1.0\,g\ cm^{-3}$ and highly compacted soil values exceeding $2\,g\ cm^{-3}$. Many arable soils have values up to $1.6\,g\ cm^{-3}$.

Urban soils that have been thoroughly cultivated such as those of allotments and flower beds have bulk densities within the range of 1.0–1.6 g cm^{-3}. The "ideal" soil for plant growth ranges from 1.45 g cm^{-3} for clays to 1.85 g cm^{-3} in loamy sands (Brady & Weil, 2002). Increasing the areas of compaction around trees reduces growth by around 50% partially due to shorter and thicker roots.

Short et al. (1986b) measured bulk densities of 1.25–1.85 g cm^{-3} (mean = 1.61 g cm^{-3}) of the surface horizon and bulk densities of 1.4–2.3 g cm^{-3} (mean = g cm^{-3}) at 0.3 m depth for open parkland in the Mall of Washington, DC. Craul (1985) found values of 1.52–1.96 g cm^{-3} for subsoils in Central Park, New York City. Patterson (1976) found average values in Washington, DC, ranging from 1.74 to 2.18 g cm^{-3}. These are large values in comparison to most grassland areas, but comparable values may well be found in many urban parks (Bullock & Gregory, 1991). Hiller (2000) found that soils in abandoned shunting yards in the Ruhr area, Germany, had bulk densities ranging from 1.0 to 2.1 g cm^{-3} depending upon the site and depth of soil where the measurements was taken. Typically, in all soils, the highest levels of bulk density occurred between 10 and 30 cm below the surface. Within this depth, bulk density values ranged from 1.6–2.1 g cm^{-3}. These were typically higher than those of a nearby wheat field, which had bulk density values from 1.33 to 1.57 g cm^{-3}.

In developing countries, Jim (1998b) found similar ranges, 1.6–1.8 g cm^{-3} for parks in Hong Kong, a tropical site. Jim (1998a) also found an average bulk density of 1.66 g cm^{-3} for 50 street side soil samples in Hong Kong. In the tropics, bulk densities range from 1.1 to 1.4 g cm^{-3} (Jim, 1998b). A recent study in Ibadan suggests that bulk density of urban soils range from 1.05 to 2.18 g cm^{-3}, with a mean of 1.62 g cm^{-3}. While these values are lower than those found in developed countries they are still statistically different (and higher than) those values for soil bulk densities found outside Ibadan in agricultural land and from those found in suburban zones (Gbadegesin & Olabode, 2000).

Poaching, an extreme form of compaction, is often seen in cities around narrow pathways and entrances wherever people or machinery are forced to tread repeatedly or track the soil at a water content close to field capacity (Bullock & Gregory, 1991).

10.3.1.3 Surface Crusting and Water Repellency

During development, soils develop a crumb structure in which particles aggregate together, thus increasing bulk volume by the presence of large pore spaces between the aggregates. This type of structure has a favorable effect on aeration, water permeability, and root penetration. In contrast, in urban settings, a thin crust appears on bare soil.

Soil surface crusting is caused by several factors, the most important being compaction of the surface layer by foot and light wheel traffic and associated lack of vegetative groundcover. The absence of the groundcover eliminates the cushioning effect of the plant shoots, the binding and lightening effect of the root system, and the contribution of organic matter to the soil surface. Thereafter, the force of raindrop splash disintegrates soil aggregates and the dispersed particles of very fine sand, silt, and clay fill the adjacent pores (Hillel, 1980). A second cause of crusting is chemical dispersion enhanced by the low electrolyte concentration of rainwater. The result is one or two micro-layers of horizontally oriented particles in the upper 2 cm of soil with an upper thin skin seal (0.1 mm thick) with an infill of fine particles about 2 mm thick. Compaction and infill of fine particles reduce water infiltration and gaseous exchange of oxygen and carbon dioxide between the atmosphere and the soil. A third possible cause is the atmospheric deposition of petroleum-base aerosols and particulates on the soil surface, which form waxy and oily substances that are water repellant. The resultant hydrophobic nature of such crust can be observed at the initiation of a light rain (Craul, 1992; Pouyat et al., 1997).

10.3.1.4   Restricted Aeration and Drainage

Urban soils are relatively confined, as they are covered with nonporous material. Where they do emerge, they are subject to crusting and compaction. Moreover, the urban soil body is often spatially interrupted by walls, sidewalks, curbs, pipe shafts, streets paths, repeated construction disturbances, foundations, or sharp change in grade. Urban soils lack the horizontal continuity between bodies present in the natural soil mantle. Together these conditions restrict gaseous diffusion and water movement in the soil profile (Craul, 1992).

The results of these restrictions are twofold. First, urban settings experience a shift from the natural local hydrological flows. Covered areas produce intensive runoff (Douglas, 1983). Bare soils, as a result of crusting and compaction, do not absorb water as quickly. As mentioned, they also dry more slowly. Covered soil may become moist and will remain so for prolonged periods of time. Lateral movements of water may be slow from effects of compaction and physical barriers. Soil conditions therefore do not present optimal conditions for plant growth. Special trees, such as the London plane tree thrive, because of their ability to cope with restriction on root space, but most street trees suffer from structural or physiological problems (Jim, 1989).

10.3.1.5   Modified Climate Regimes

The climate and hydrology of cities are different from that of rural areas and these differences impact the characteristics of soils.

The incoming radiation may be less in the city because of increased haze and cloudiness, but the amount of heat absorbed and reradiated by building and street surfaces is greater than vegetation, raising both daytime and nighttime urban air temperatures (Landsberg, 1981). The city heat island is typically conterminous with the built-up area of an urban center although wind direction can change temperatures of other areas temporarily. Urban temperature excess results in a bubble of maximum temperature different at the surface trailing off to no difference somewhere between 300 and 500 m above the city.

Increases in urban temperatures offset the latitudinal gradients of animal population densities and cycles. In Europe, various arthropod populations found in cities at
more northern latitudes were more similar to those in rural settings at more southern
latitudes than rural populations further north. A similar example of how urbanization
can counteract latitude has been noted for bird diversity. Increased heat translates
into longer growing seasons for plants and earlier seasonal blossoming.

Changes in temperature and moisture impact soil characteristics (Bullock &
Gregory, 1991). Urban soils that are surrounded by large capacity heat-absorbing
and re-radiation surfaces interact differently with flora and fauna, than those otherwise. City soils can be much warmer than those in their respective regions. Given
that moisture is not limiting, warmer temperatures increase the metabolic rate of
microbes and invertebrates, thus increasing decomposition (Craul, 1992). Pouyat
et al. (1997) found that along an urban–rural gradient in New York City, decomposition of litter occurred almost twice as fast in the urban forests than it did in rural
forests. At the same time, forest-floor litter mass by weight was almost three times
greater in the rural stand than in the urban stands, with suburban stand intermediate
(Kostel-Huges et al., 1998). While there are a number of other likely contributors
to these differences, heat island is an important one.

Moreover, soil-weathering process may be intensified. Soil surfaces form gradients where surface heat builds (Hillel, 1980). This is because the greatest temperature extremes occur in the surface horizons. Soil temperature is highly influenced by
moisture content because water affects the heat capacity of the soil. High temperatures dry the soil surface through evaporation, creating moisture gradients that cause
water to flow from lower to upper horizons (Craul, 1992). Hence water movement
within heat-affected soils is increased, arguably impacting translocation processes.

10.3.2 Urban Soil Topologies

Of course, the six characteristics described in Section 10.3.1.2 are not the only
physical differences between urban and nonurban soils. For example, Hiller (2000)
found that soils in former railway shunting yards had more magnetic susceptibility
than those of rural areas, largely due to the large amount of railway slag and ballast
in soil layers. The six impacts, however, are the major impacts that can be seen
across almost all urban soils.

These various differences encourage soil scientist to categorize urban soils differently from "natural" soils. At the same time, given the nature of urban soils, categorizing them into traditional taxonomic groups is difficult (Short et al., 1986a). In England
and Wales, one of the 10 major groups recognized by the Soil Survey is that of human-
made soils defined as soils with a thick human-made A horizon, a disturbed subsurface
layer, or both. Gilbert (1989) has studied urban sites there and suggests they produce
lithomorphic soils (i.e., A/C soils with no B horizons) which would be classified by
the Soil Survey of England as pararendzinas. Bullock and Gregory (1991), on the other
hand, consider urban soils "Anthroposols" which can then be further subdivided into

suborders such as Cumulic, Hortic, Garbic, Urbic, Dredgic, Spolic, and Scalpic, depending upon the exact form in which the original soil was modified, mixed, truncated, or buried or what new soil parent materials were added.[9]

In the USA, Craul (1992) suggests that urban soil profiles are generally classed as Entisols, mainly the Udorthents, having little horizon differentiation, exhibiting an absence of diagnostic epipedons in the disturbed portion (anthropeic horizon) and having a udic moisture regime. There may be diagnostic subsoil horizons below the anthropeic horizon within the truncated original profile, or there may be fragments of diagnostic horizons mixed in the anthropeic horizon.

In Germany, Blume (1989) suggested that urban soils be considered "deposit soils" or Urbic Anthrosols. They are further subdivided by natural processes of soil genesis. Accordingly, these soils are initially raw (Syrosems) that developed into Regosols, Rankers, Pararendzinas, or Rendzinas (Hiller, 2000).

The classification of urban soils as Urban Anthrosols has been further promoted by Food and Agricultural Organization (FAO) and United Nations Educational, Scientific, and Cultural Organization (UNESCO). Currently, the FAO–UNESCO world soil map describes Anthrosols as formed or profoundly modified through human activities such as addition of organic materials or household wastes, irrigation, or cultivation. The group includes soils otherwise known as "Plaggen soils", "Paddy soils", "Oasis soils", and "Terra Preta do Indio". Anthrosols are differentiated by their anthraquic, hortic (formed in layers from kitchen refuse, oyster shells, fishbones, etc., from early Indian habitation), hydragric (formed by long-continued wet cultivation such as "puddling" of wetland rice fields), irragric (formed from prolonged sedimentation of silt from irrigation water), terric (addition of calcareous soil materials, e.g., beach sands) or plaggic (produced by long-continued addition of "pot stable" bedding material, a mixture of organic manure and earth) surface horizon (Driessen et al., 2001).

10.3.3   Regional- and Global-Scale Impacts of Urbanization on Soils

Urban areas have reached extremely large sizes, both in the developed and developing worlds. Amongst a larger set of regional and global issues, some important soil impacts include loss of agricultural, wetland, and other soils to urban land uses and the acceleration of erosion at the local, citywide, and even global level. These impacts have indirect effects on soils including soil loss, sealing, and degradation. On the other hand, in some countries, rural–urban migration is associated with forest regeneration and potential for soil recovery. Moreover, in some poor nations, difficult access to cities can increase soil quality (Braimoh & Vlek, 2006).

---

[9] See also more detailed classifications given by Australian soil categorization of "Anthroposols" at http://www.clw.csiro.au/aclep/asc_re_on_line/an/anthsols.htm (accessed on 10 September 2006).

10.3.3.1 Loss of Agricultural, Wetland, and Other Soils Due to Urbanization

In general, land-use change from nonurban to urban land uses reduces the total net primary productivity (NPP) of Earth as urban land reduces biomass quantity. This human appropriation of NPP impacts natural ecosystems and the organism that comprise them, leading in the long run to the loss of biodiversity (Hassan et al., 2005). The removal of agricultural lands by urban land uses has been of concern, not only because of appropriation of NPP, but also because of the perceived constraints it places on future increases in food production, national food security, and environmental sustainability (see, e.g., Ehrlich, 1991; Rerat & Kaushik, 1995; Gardner, 1996). Between 1950 and 2000, acreage of land planted in grains per person worldwide shrunk by half. Nearly all the 2.6-fold increase in grain production since 1950 has come from increasing yields (production per hectare). While on the average there is more food produced per person today than 50 years ago, since 1984, grain production per capita has fallen as harvest increases have failed to keep pace with population growth (Ehrlich, 1991). World population growth is outpacing the expansion of cropland (see Chapter 3).

As the world urbanizes, agricultural land is transformed to urban land. This is because cities have traditional been established around areas of high ecosystem service production. Urban sprawl, as the process of urban land use expansion is sometimes called, is of increasing concern around the world. Globally, urban land-use urban activities potentially removes 6.8% of all cultivated systems from agricultural production (McGranahan et al., 2005). This suggests the removal of at least 2.4 million km² globally from agricultural production, of which a proportion is high quality farmland.

Within the developed world, such as the USA, there has been increasing public concern over the long-term impact of urbanization on the available agricultural land. Until recently, however, it was difficult to measure and analyze accurately the trends in urban expansion and their impacts on agricultural lands. Certainly over the years estimates of agricultural land loss have varied dramatically (Hart, 1976).

In 1956, Bogue picked up on growing concern over the growth of cities in the USA and calculated that over 150,000 ha of cropland and pastureland would be converted to urban uses each year between 1955 and 1975 in metropolitan counties alone (Bogue, 1956). In the same year, D.A. Williams, administrator of the Soil Conservation Service, proclaimed that over 455,000 ha of cultivable land were being converted to urban use each year in the whole country. This figure became a doctrine in the Soil Conservation Service.

In the 1970s, two events helped to keep urbanization's impacts on agriculture lands in the forefront of popular attention. First, due to high demand from the Former Soviet Union, there was a dramatic rise in world grain prices. US sales of grain eliminated the excess capacity that characterized US agriculture since World War II. Second, studies by the US Soil Conservation Service indicated that urban land was rapidly expanding in the nation. This was due, in large part to the suburbanization of large numbers of middle-class Americans leaving central city areas and that most of the residential constructions occurred on formerly agricultural lands (Fischel, 1982). Reports suggested that from 1958 to 1967, the average annual conversion from agriculture to built-up and urban land was 460,000 ha, from 1967 to 1977 the average annual amount was more than 1.1 million ha and from 1967 to 1975 the annual amount was 842,000 ha (Fischel, 1982). Subsequent estimates place the annual conversion of land at more than 2,000 ha year[-1] (Hart, 1976). These national studies were supplemented by regional studies of fast-growing counties. Berry (1978) presented data summarizing three such studies that argued urban land conversion ate up between 2.1% and 3.9% of agricultural land in their areas. Between 1960 and 1980, urban land in the USA increased by between 9 million ha and 12 million ha, most of this taken from cropland, pastures, and forests (Frey, 1984). Of this amount, it was estimated that 2 million ha were prime agricultural land. The categorization of "prime" farmland was based upon the ability of the land to produce food, feed, forage, fiber, and oilseed crops.

Not everyone agreed with these estimates, however, Some studies indicated that annual loss of agricultural land to urban built-up areas during 1967–1977 was only 1.1 million ha, which was less than 1% of the total 132 million ha available land at the time (Plaut, 1980). By the early 1990s, studies began appearing that discounted any threat to food and fiber production capacity from continued urbanization (Vesterby & Heimlich, 1991). Analysis of urban land conversions processes in the 1960s, 1970s, and early 1980s demonstrated that consumption rates are constant, rather than rapidly increasing, during these periods (Vesterby & Heimlich, 1991).

Within the fast-growing areas, total urban land was 3.6 million ha in the early 1980s, an increase of 0.97 million ha or 37%. The residential component of urban land-use change increased by more than 50% from the early 1970s to the early 1980s. Agricultural land contributed about one-third, with rangeland making up about 61% of this. While during the 1970s, the USA experienced a "rural renaissance," with rapid population growth in nonmetropolitan areas where marginal land consumption was greater than in metro areas, population growth in the 1980s returned to the pre-1970s patterns of slower growth outside metropolitan areas (Vesterby & Heimlich, 1991).

In the late 1990s, US Department of Agriculture studies reevaluated the current state and trends in land uses throughout the country. These estimates suggested that US land in urban uses (for homes, schools, office buildings, shopping sites, and other commercial and industrial uses) increased 285% from 6 million acres in 1945 to an estimated 23 million acres in 1992. These totals were way below the 1970s predictions that urban land was taking 0.45–1.13 million ha of agricultural land a year. The estimates, however, imply that urban land use grew faster than population. While the US population nearly doubled, the amount of land urbanized almost quadrupled. However, urban uses still amount to only 3% of total national land area. At the same time, land in transportation uses (highways and roads, and airports in rural areas) increased by 1.6 million ha (17%) between 1945 and 1982. Transportation uses declined by 0.8 million ha from 1982 to 1992 due to the abandonment of railroad facilities and rural roads, and the conversion of some land under transportation uses into urban uses. From 1992 to 1997, urban land use increased by 6%. Land in urban areas is estimated at 27 million ha in 1997. Rural residential land, a new land-use category, was estimated to be about 30 million ha in 1997 (Vesterby & Krupa, 2001).

Cropland in the contiguous 48 states of USA decreased by about 26% from 1945 to 1992. At the same time, however, cropland idled and pasture increased by 15% and 19%, respectively. As a result, total cropland (used for crops, used for pasture, and idled) has decreased only slightly since the late 1960s. The fall in areas devoted to crops was largely due to changes in cropland idled in federal crop programs and weather, such as the drought in 1988 and the heavy rains in 1993 (USDA 1997). Between 1992 and 1997, all cropland in the contiguous 48 states of USA decreased by 2 million ha (1%). This follows a slow but steady decline since 1978 by about 3%. During this period, however, areas devoted to crops increased by more than 4 million ha (Vesterby & Krupa, 2001).

A recent estimate of farmland lost was reported through the use of satellite imagery (Nizeyimana et al., 2001). In this study, agricultural lands were categorized into four soil productivity classes, namely high, moderately high, moderate, and low. Within the USA, the distribution was 3%, 26%, 38%, and 33%, respectively. Urban land uses were distributed as 6% on high, 48% on moderately high, 35% on moderate, and 11% on low productivity soils. Although the land with the most productive soils represents a small fraction of the total land area in several states, this category experienced the highest level of urbanization. The level of urbanization were estimated as 1%, 2%, 4%, and 5% in the low, moderate, moderately high, and high soil productivity classes, respectively (Nizeyimana et al., 2001).

The most recent United States Department of Agriculture farmland assessment suggests that urban land areas quadrupled from 1945 to 2002, increasing about twice the rate of population growth over this period. After adjusting earlier estimates for new criteria used in the 2000 Census, urban area increased by 13% between 1990 and 2002. Census estimates based on the previous criteria indicate that urban area increased 9 million acres (18%) over the 1980s, 13 million acres (37%) over the 1970s, and 9 million acres (36%) over the 1960s. Estimated area of rural land used for residential purposes increased by 9 million ha (29%) from 1997 to 2002, and by 7 million ha (30%) from 1980 to 1997 (Lubowski et al., 2006).

This history is instructive in that it demonstrates the controversy over estimates. First, the conversion of rural to urban land is far from straightforward. Second, it is only recently with the development of new technologies to study land-use change that we are able to more accurately monitor, analyze, and understand the trends of land conversion. Third, while some previous estimates appear to be exaggerated, current more conservative estimates are also worrisome. As best as understood, urban land-use conversion of agricultural land stood at 455,000 ha per year between 1958 and 1967 and 570,000 ha for the period 1967–1975 (Vesterby & Krupa, 2001). Between 1992 and 1997, land-use conversion to urban continued at a rate of 570,000 ha per year (Vesterby & Krupa, 2001), with high productivity farmland as the most pressured (Nizeyimana et al., 2001). The new category of "rural residential" includes more area than the total amount of urban land (Lubowski et al., 2006).

In Europe, cities tend to be more compact than those of the USA. Since the 1950s, however, they have been growing and sprawl is now seen as a threat on soils. For example, since 1950s, European cities expanded on average by 78% while the population has grown by only 33%. In some places, such as Spain, Portugal, and parts of Italy, population is decreasing while urban areas are still growing. During the period 1990–2000, the growth of urban areas throughout Europe consumed more than 8,000 km². This is equivalent to the complete coverage of the state of Luxembourg or the consumption of 0.25% of the combined area of agriculture, forest, and natural land on the continent. While these increases may seem small, a 0.6% annual increase in urban areas would translate into a doubling of the current area in a little over a century (European Environment Agency, 2006).

The loss of agricultural and wetland to urbanization is considered severe in the developing countries (Douglas, 1994). In many cases, fear of agricultural land loss is compounded by food shortages where population increase and urbanization is occurring fastest. In China, Thailand, India, Indonesia, and Vietnam, for example, urbanization is responsible for decreasing cropland area (Gardner, 1996). From 1978 to 2002, Chengdu, China, increased by about three times, from 81 km² to nearly 220 km². Some of the land overtaken was previously agricultural, much of it green space (Schneider et al., 2003). The growth of Mexico City took 53,000 ha of agricultural land from 1960 to 1980 (FAO, 1985). From 1966 to 1971, 14,000 ha of irrigated land were lost to urban growth around Lima, Peru (Blitzer et al., 1981). India lost approximately 1.5 million ha of land to urban growth from 1955 to 1985 (Chhabra, 1985). From 1987 to 2001, the urban land area of Istanbul, Turkey, increased by about 15,000 ha (despite an earthquake during this period that led to loss of urban land) which translated into a 127% increase. Meanwhile the population increased only by 83% between 1985 and 2000. In contrast, forest, seminatural vegetation, cropland, and bare soil areas decreased, particularly those areas close to the city (Kaya & Curran, 2006). In the Philippines, between 1988 and 1995, the country lost approximately 33,700 ha of agricultural land, half of which was transformed to nonagricultural uses around Metro Manila (Kelly, 1998).

Agricultural land is not the only type of land use encroached by cities. Urban expansion typically penetrates other sensitive environments. Most cities, because they had been founded close to inland waterways or on coastal zones, have expanded into their floodplains, narrowing the channels available for the passage of storm water and engulfing sensitive coastal areas. Approximately, 58% of the world's major reefs occur within 50 km of major urban centers of 100,000 people or more, while 64% of all mangrove forest and 62% of all major estuaries occur near such center (Agardy et al., 2005b). This trend is increasing as giant coastal cities have emerged in many parts of the world. Indeed, coastal populations on every continent have grown with global trade into cities with international ports, creating jobs, and economic growth. Urbanization and all that it entails is an important driving force in coastal zone change. Toronto, Canada, for example, expanded its industrial, utility, and residential activities onto marshland from 1931 to 1976, reclaiming 271 of the 482 ha of marches (Lemay & Mulamootil, 1984). In the USA, most of the loss of wetlands occurred prior to the onset of rapid urbanization during the colonial period and early history of the country's development (USDA, 1997). Urban development impacts natural ecosystems, productive wetlands, and habitats of wildlife including threatened and endangered species. Wetlands loss in the USA continues but at a much slower rate than in previous decades. The National Resource Inventory estimates that half of the over 40,000 ha of wetlands lost per year from 1992 to 1997 was to urban land development (USDA, 2001). In most cases new wetlands from restoration and creation are of lower quality than lost wetlands (Randolph, 2004). The latest wetlands status and trends report by the US Department of the Interior estimate annual net loss at about 24,000 ha per year over the 1986–1997 period (USDI, 2001). This does not yet conform to the federal "no-net-loss" policy, but it is a substantial improvement over estimates of loss in previous decades (Randolph, 2004)

In the developing world, the loss of valuable ecosystems, due to urbanization and to satisfy urban demands, is also reaching critical dimensions. For example, more than 40% of the world's estimated 18 million ha of mangrove forest occur in South and Southeast Asia (UN ESCAP, 2000). Large areas of mangrove have been removed for industrial, residential, and recreation developments, and in particular for establishment of ponds for fish and prawns aquaculture. Within Southeast Asia, the loss of original mangrove areas is high; less than 20% in Brunei, more than 30% in Malaysia, more than 40% in Indonesia, more than 55% in Myanmar, more than 60% in Vietnam, more than 70% in Singapore, and more than 80% in Thailand (UN ESCAP, 2000). In the Philippines, 210,500 ha of mangrove (approximately 40% of the country's mangrove cover) were lost to aquaculture from 1918 to 1988. By 1993, only 123,000 ha of mangroves were left, equivalent to a loss of 70% (Agardy et al., 2005a).

The conversion of natural or agricultural landscapes to highly modified urban landscapes throughout the world is expected to continue. This trend is particularly important to developing countries. This is because, as mentioned previously, almost all global population growth will occur in the developing world cities (United Nations, 2006). Many areas, in both the developed and developing world are expected to become even more highly modified than they are today.

10.3.3.2   Accelerated Erosion at all Scales

No soil phenomenon is more destructive worldwide than erosion, the detachment and movement of soil or rock by water, wind, ice, or gravity. While erosion processes have always been part of the geologic cycle and human history, accelerated erosion which is more rapid than normal, natural, geological erosion, resulting largely from human impact is more ominous now than any time in history (Brady & Weil, 2002; see also Chapters 4 and 5).

Humans move increasingly large amounts of rock and sediment in the cause of urban development. Simultaneously, as digging sticks and antlers have given way to wooden plows, iron spades, steam shovels, and today's huge excavators, our ability and motivation to modify the landscape by moving earth in construction and mining activities have also increased dramatically. In particular, since the development of the bulldozer in the 1950s, the soil cover of building sites, for example, has been increasingly disturbed. Instead of constructing houses and other infrastructure in line with the contours of a given site, the bulldozer has facilitated construction so that some previously sloping sites contain soil profiles beneath bulldozed soils and parent material from further upslope (Bullock and Gregory, 1991). As a consequence of this and other earthmoving activities, humans have now become arguably the premier geomorphic agent sculpting the landscape, and the rate at which we are moving earth is increasing exponentially (Hooke, 2000).

Moving large amounts of earth and soil, however, has also led to increased erosion. The current global mean rate of soil loss may exceed rates of formation by up to an order of magnitude. That is, over the last half-billion years of the Earth's history, erosion has lowered continental surfaces by a few tens of meters per million years. In comparison, construction and agricultural activities currently result in the transport of enough sediment and rock to lower continental surfaces by a few hundred meters per million years (Wilkinson, 2005). During the last 40 years, nearly one-third of the world's arable land has been lost by erosion and continues to be lost at a rate of more than 10 million ha year^{-1}. The on-site costs[10] are between US$4 billion and US$27 billion, while the off-site costs[11] could be higher (from US$5 billion to US$17 billion) (Pimentel et al., 1995). In the USA alone, the estimated loss of total readily available nitrogen, phosphorus, and potassium are in the order of 9,494 Mg total nitrogen, 1,744 Mg available nitrogen, 1,704 Mg total phosphorus, 34.1 Mg available phosphorus, 57,920 Mg total potassium, and 1,158 Mg available potassium (Larson et al., 1983).

While the movement of rock and soil is largely the result of agricultural processes, a significant proportion, up to 30%, is from construction and activities related to urbanization (Hooke, 2000). Urbanization increases runoff pollution, erosion, and sedimentation through land clearing and increased impervious surfaces. Land activities that remove or reduce vegetative cover and litter, loosen or compact the surface soil from its natural state, create channelization of surface runoff, or interrupt natural drainage increase the soil-erosion potential.

Runoff pollution results in unavoidable degradation at 30% impervious surfaces (Schueler, 1994). Moreover, impervious surfaces prevent natural pollutant processes in soils by preventing percolation. In the USA, nonpoint source runoff is the largest sources of surface water pollution (Randolph, 2004) and impervious surface coverage is emerging as a key environmental indicator of this challenge.

---

[10] On-site damages include loss of soil and nutrients, spread of plant-disease organism creation of dense crust on the soil surface (which reduces water infiltration and increases water runoff), difficultly with establishment of plants as seed and seedlings are washed away, and that gullies care up badly eroded land affecting building foundations, and causing unsafe conditions.

[11] Off-site costs include damages from sediments to other crops, vegetation, and infrastructure; damages from windblown dusts to crops, buildings, and infrastructure; reduction of recreational space and its value; water pollution and increased human morbidity and mortality from windblown dusts; and dispersal of waterborne diseases.

Rickson (2003) states that mined catchments have 69 times more sediment per unit area than unmined catchments, that unrestored slope banks have 968 times more sediment than undisturbed areas, and that haul roads associated with mining activity produce 2,065 times more sediment in the surrounding slopes. Douglas (1986) suggests that urbanization (new residential, industrial, or commercial development) often leads to erosion of surface soil which increases silt loads of streams by as much as 50–100 times. Erosion increases due to exposure of highly erodible material, lack of cover and slope characteristics (as affected by construction).

Data on soil loss from various land uses (Hough, 2004) shows that the largest loss is from exposed construction sites. These types of developments can loose three orders of magnitude of soil than forest land (almost 93,000 kg ha^{-1} year^{-1}) (Hough, 2004). While erosion from newly developed land may be massive, these are initial pulses and do not continue, even if land remains undeveloped. However, where construction occurs, or the vegetation is removed, the erosion rate is great. In 1978, 80 million t of soils eroded from construction sites and 169 million t from roads and roadsides, and nearly 90% of this takes place on land under development (Craul, 1992).

Cultivated agricultural land loses more soil than developing and developed urban areas. Erosion rates in developed urban areas, however, can easily be twice those of forested areas (Hough, 2004). The erosion rate in developed urban areas, which has established vegetation or consists of noneroding surfaces are not great. The erosion that does occur in these circumstances typically comes from stream bank erosion in the remaining unchannelized creeks and streams in the city.

While erosion rates differ by climate, geology, geomorphology, and extent of human impact, they are considerable in both developed and developing world cities. In a study of erosion in developing humid tropical urban areas, Douglas (1978) concluded that urban and mining activities greatly accelerated erosion by leaving areas of soil less protected than under plant cover in an entire catchment area. Douglas (1974) compared sedimentation rates within watersheds of developed (Armidale, New Zealand) and developing (Kuala Lumpur, Malaysia) urban catchments. He found in both areas, small tributaries are significantly impacted through increased sedimentation (through erosion) by construction work. This has changed the hydrologic regimes within both these cities, although it was particularly acute in Kuala Lumpur, through loss of channel capacities, among other changes. Hardoy et al. (2001) have identified erosion as one of the most important environmental issues in low-income urban areas.[12]

Within developed countries, such as the USA, erosion has had significant impacts on cities. Over the next 60 years, for example, erosion rates on the east

---

[12] In the tropics, land degradation occurs sooner after human impact and is longer lasting than in the temperate zone. Especially in the moist tropics, a large portion of nutrients are contained in the vegetation of tropical forests, which often grow on thin, poor soils that are subject to erosion, leaching, and compaction when deforested. Deforestation and conventional (temperate zone) agriculture frequently has disastrous results (Ehrlich, 1991).

coast (if they continue their current levels of at least 60–90 cm year^{-1}) will threaten 53,000 existing and 23,000 currently planned structures (Randolph, 2004).

Gully erosion, whereby water accumulates in narrow channels and over short periods removes the soil from this narrow area to considerable depths, ranging from 0.5 to 30 m (Brady & Weil, 2002), has been ascribed to both agricultural and pastoral activities as well as road building and urbanization (Rickson, 2003). Roads induce a concentration of surface runoff, a diversion of concentrated runoff to other catchments, and an increase in catchment size, which enhance gully development after road building. Also changes in drainage patterns associated with urbanization can result in gullying. Gullies often take place where there are illegal settlement without urban infrastructure, such as sanitation and paved roads. This is because these types of settlements emerge on steep unprotected slopes.

10.3.3.3   Urbanization and Soil Biology

Soil is one of the most diverse ecosystems on earth. It comprises several thousand individual organisms including fungi, nematodes, protozoa, and bacteria. The composition of these organisms varies with organic material, moisture, and clay content of the soil. Urban soils generally have lower biological capacity compared to agricultural soils due to the highly variable environmental conditions of urban soils. The soil–microbe complex is characterized by highly ordered heterogeneity and the spatial clustering of the matrix results in the characteristics of the soil physical structure. The soil physical structure in turn provides habitat that enables several biophysical and biochemical processes that regulate life on Earth. Such ecosystem services include plant productivity, water retention, buffering capacity to water-bound pollutants, and greenhouse gas emissions (Young & Crawford, 2004).

Urban development introduces marked changes along the rural–urban gradient. Increases in anthropogenic disturbance often results in habitat loss, leading to considerable local extinction rates, elimination of large majority of native species, and replacement of native species with widespread nonnative species (McKinney, 2002). Such a process of biotic homogenization constitutes a threat to the biological uniqueness of ecosystems. Several studies document that the lowest species diversity including earthworm and microbial biomass along the rural–urban gradient occur in the urban core. Reduction in richness is often associated with loss of vegetation and the pollutants (especially heavy metals) introduced to urban environments.

Because urban soils tend to be highly modified in comparison to natural soils, they have the tendency to be more structurally and functionally altered. Soil nematodes regulate microbial populations and nutrient cycling in the soils. They can live in wide-ranging environmental conditions and occupy several positions at the trophic levels. In particular, they are an important indicator of soil quality and human disturbance on soils (Pavao-Zuckerman & Coleman, 2007). Studies on nematode fauna along rural–urban gradients have produced mixed results. Pavao-Zuckerman and Coleman (2007) observe that the diversity of nematode genera was not affected by the physico-chemical conditions associated with urban land use in Asheville, North Carolina. In contrast, functional differences were observed along land-use gradient, with urban soils having lower abundance of predatory and omnivores nematodes. Furthermore, less fungal dominance in the soil food webs of urban soils was observed. These results differ markedly from that of Pouyat et al. (1994) who observed lower density of nematodes in urban compared to rural soils in a rural–urban gradient in Bronx, New York. Pouyat et al. (1994) also observed significantly higher fungal abundance in rural compared to urban soils. They further observed an inverse correlation between soil heavy metal concentration and both fungivorous micro-invertebrates and litter fungi.

The impacts of urbanization on soil enzyme activities have been studied recently. Green and Oleksyszyn (2002) measured the activities of invertase, cellulose, and CO_2 flux in mesiscape (irrigated water lawns), xeriscapes (low-water-use vegetation), and desert remnants (undeveloped areas within the urban matrix) in central Arizona. Differences of invertase activities between land uses were most pronounced in the winter months, when value for mesiscape sites was about twice as much as the desert remnant sites, and over three times greater than the xeriscape sites. Invertase activity was significantly higher in desert remnant sites compared to xeriscape sites only in December and January. In January, cellulase activity in mesiscape soils significantly exceeded xeriscape and desert remnant soils by a factor of 2.

Microbial processes of urban and rural soils have also been compared by some researchers. There were few differences in denitrification potential (denitrification enzyme activity) between urban and rural and herbaceous and forest riparian zones, though higher variability was observed in urban soils (Goffman & Crawford, 2003). Similarly, no significant differences were observed in microbial biomass. However, denitrification potential was strongly correlated with soil moisture and organic matter content. This suggests that if surface runoff can be channeled through areas with high denitrification potential such as wetland basins, they could function as important nitrate sinks in urban watersheds (Goffman & Crawford, 2003).

10.3.3.4   Urbanization and Reforestation in the Developed World

The increase in the proportion of urban population relative to total population can translate into the return of land previously used for agricultural purposes to forested land in some parts of the world. For example, European urbanization levels have reached 76% and at the same time, population growth for the continent is very low (<0.03% per year), if not already negative (United Nations, 2006). Contemporary population growth in, and expansion of, cities results in losses in population and area in rural parts of the continent. It is therefore not surprising that in Europe forest areas are actually increasing in range. For all of Europe, from 1990 to 2000, forests increased in area by 0.09% and from 2000 to –2005, forests increased in area by 0.07%. Some of the biggest gains were found in countries such as Finland, France, Greece, Denmark, Iceland, Ireland, Portugal, Italy, Spain, and Switzerland (FAO,

2006). The expansion of forests on abandoned agricultural land is quite common in some parts of Europe. Combined with an emphasis away from productive functions toward conservation of biological diversity, protection and multiple uses, and the increase in forest area can be associated with increases in soil quality. This trend (urbanization leading to abandoned and reforested farmland with positive consequences for soil quality) is also true, but to a lesser extent in other parts of the developed world, such as North America and Japan (FAO, 2006).

10.3.3.5 Urbanization, Access to Markets, and Increases in Soil Quality in the Developing World

While there are studies that suggest urban impacts are always negative on land outside cities, some evidence suggests that urbanization can improve soil quality (e.g., Jacoby, 2000). Specifically, in low-income countries where rural farmers have access to markets in towns and cities, they also have access to soil supplements including fertilizers. Applications of fertilizers to soils in these regions typically improve soil nutrient quality and hence have a positive impact on soil quality.

10.4   Urbanization, Soil pH, and Heavy Metal Contamination

10.4.1   Urbanization and Soil pH

Urban soils tend to have soil reaction (pH) values somewhat different from their natural counterparts. In most cases, pH values are higher, more alkaline, in the urban environment (Craul & Klein, 1980).

There are several explanations for the elevated pH values of urban soils. First is the application of calcium or sodium chlorides as sidewalk and street-deicing salts in northern cities, as in Syracuse (Craul, 1992). The irrigation of urban soils with calcium-enriched water is the second cause. The third explanation is atmospheric pollution. Ash residue often contains calcium. This is further evidenced in the decrease in pH values with increasing depth. The fourth explanation relates to the release of calcium from weathering of construction rubble comprised of bricks, cement, plaster, etc., or the washing of calcium from building facades by polluted acid precipitation. Finally, liming of soil to correct suspected deficiencies raises the pH. Gilbert (1989) has stressed on the chemical properties of rubble soils as the source of high pH in the UK. Brick rubble contains calcium, important as a nutrient and controller of pH, which maintains soil pH in the range 6.5–8.0.

In the Mall in Washington, DC, fill materials contaminated with building rubble are acidic, with pH of the horizons varying from 6.4 in the surface to 6.7 in the fifth horizon (Short et al., 1986b). In Central Park, New York City, fill material also has an acid pH (Warner & Hanna, 1982). In the UK, local roadside and path-side soils have pH around 9.0 as a result of sodium and calcium chloride applied as deicing salt (Gilbert, 1989). A study of the railroad shunting areas in Ruhr region, Germany, however, suggested that pH levels were lower than those of agricultural land (Hiller, 2000). Sukopp et al. (1979) report high values for pH of the soils in city streets in Berlin (8.0), which decrease as one moves to city forests. Mean pH values for five different land uses (agriculture, ornamental gardens, parks, riverbanks, and roadsides) in Seville was approximately 7.21 (Ruiz-Cortes et al., 2005). Soils in Palermo, Italy, exhibited a narrow alkaline range of pH (7.2–8.3) (Manta et al., 2002). Soil samples in the western district of Naples, Italy, had average pH levels of 8.01, while those in the eastern district averaged 7.4 and the central district samples averaged 6.8 (Imperato et al., 2003). The alkaline pH levels relate to the presence of carbonates in soil, while in the central urban district, some pH values fell lower than 6.0. Garden soils in Salamanca, Spain, also demonstrated lower pH values, between 3.4 and 7.6 (mean 6.7) (Sanchez-Camazano et al., 1994).

Within developing world cities, Jim (1998b) found alkaline humid tropical soil in Hong Kong with pH exceeding 8.0 in some cases. In Caracas, Venezuela, soils along roadsides polluted had pH levels of between 7.5 and 7.8 (Garcia-Miragaya et al., 1981). Krishna and Govil (2005) found neutral to alkaline pH levels (pH 7.5–8.5) in Thane-Belapur industrial development area of Mumbai. In satellite cities of Seoul, Chon et al. (1998) found pH levels of 5.2, 5.7, 6.0, and 7.6 for soils in forests, paddy, dry fields, and roadsides, respectively. The suggested the relatively high pH values in roadside soils were due to commenting materials associated with the road surface. On the other hand, Gbadegesin and Olabode (2000) found soil pH in Ibadan to be between 5.1 and 7.6 with a mean of 6.62. These values however, were not significantly different from pH values in soils from suburban and agricultural areas. Soils studies in Bangkok also found lower pH levels in soils ranging from 3.6 to 7.4 with a mean of 6.7 (Wilcke et al., 1998).

10.4.2 Heavy Metals in Soils

Heavy metals are a subset of persistent toxics that retain their environmental impact for a relatively long period of time after release into the ecosystem. Elements in soils include Cd, Co, Cr, Cu, Fe, Hg, Mn, Mo, Pb, and Zn (Brady & Wiel, 2002). Since they are elements, they never decay. They can change to different forms that may increase or decrease their toxicity. Additionally, they can disperse, accumulate, or undergo other physical changes leading to changes in the likelihood of exposure, thus altering associated risks in the environment.

Trace metal distribution in soils is well documented for many industrialized countries such as Japan, Germany, and the USA. Both natural and human activities introduce trace metals into the environment. The presence of heavy metals, salts, and acid deposition products is primarily due to the environmental conditions created by urban areas (see Table 10.5 for examples of sources). The heavy metals are deposited on the soil surface from the atmosphere as products of fossil fuel combustion of vehicles, power plants, and industrial processes.

Anthropogenic sources include production-related activities such as use, wear, and disposal of consumer and commercial products. Among the production-related activities, the high-temperature processes such as smelting of nonferrous metals; iron production; other industrial activities; waste incineration; phosphorus fertilizer production; coal, wood, and petroleum combustion are significant contributors to metal pollution. The bulk of the arsenic and cadmium emissions comes from metal smelters, whereas selenium is primarily contributed by coal combustion. Lead is introduced mostly by combustion of leaded gasoline. Environmental mercury is introduced by a mixture of technological activities, with energy generation responsible for approximately 50% of the total.

In addition to the production-related processes, the normal use and wear of consumer and commercial products are an important source of trace element releases. The dissipative consumption includes: weathering of paints and pigments (silver, arsenic, chromium, cadmium, copper, mercury, lead, and zinc); incineration of discarded pharmaceuticals (silver, arsenic, chromium, and zinc), batteries (mercury, cadmium), electronic tubes (mercury), plastics (zinc) and photographic film (silver,); wear and weathering of electroplated surfaces (cadmium), leather (chro-

Table 10.5 Potential sources of heavy metals

Element	Source
Arsenic	Pesticides, fertilizers, plant desiccants, animal feed additives, copper smelting, sewage sludge, coal combustion, incineration and incineration ash, detergents, petroleum combustion, treated wood, mine tailings, parent rock material
Cadmium	Phosphate fertilizers, farmyard manure, industrial processes (electroplating, nonferrous metal, iron, and steel production), fossil-fuel combustion, incineration, sewage sludge, lead, and zinc smelting, mine tailings, pigments for plastics and paint residues, plastic stabilizers, batteries, parent rock material
Chromium	Fertilizers, metallurgic industries, electric arc furnaces, ferrochrome production, refractory brick production, iron and steel production, cement, sewage sludge, incineration and incineration ash, chrome-plated products, pigments, leather tanning, parent rock material
Nickel	Fertilizers, fuel and residual oil combustion, alloy manufacture, nickel mining and smelting, sewage sludge, incineration and incineration ash, electroplating, batteries, parent rock material
Copper	Fertilizers, fungicides, farmyard manures, sewage sludge, industrial processes, copper dust, incineration ash, mine tailings, parent rock material
Lead	Mining, smelting activities, farmyard manures, sewage sludge, fossil-fuel combustion, pesticides, batteries, paint pigment, solder in water pipes, steel mill residues
Manganese	Fertilizers, parent rock material
Mercury	Fertilizers, pesticides, lime, manures, sewage sludge, catalysts for synthetic polymers, metallurgy, thermometers, coal combustion, parent rock material
Zinc	Fertilizers, pesticides, coal and fossil-fuel combustion, nonferrous metal smelting, galvanized iron and steel, alloys, brass, rubber manufacture, oil tires, sewage sludge, batteries, brass, rubber production, parent rock material

mium), plastics (zinc); and decomposition or combustion of treated wood (arsenic, chromium, copper). Whereas ambient air is the principal initial target of metal emission from some of these processes, it is the soil and surface waters that receive the bulk of the emissions from dissipative consumption, mainly via surface runoff and sewage-treatment plants. The relative contributions of consumption-related and production-related emissions to the total environmental loading of metals are largely unknown (Brown et al., 1990).

Metals emitted into the ambient air from various sources are carried different distances by the winds, depending on their state (gaseous, vapor, or particulate) before they fall or are washed out of the air onto land or the surface of the oceans. Particulate matter particle size is a decisive factor. Most metal associated with coarse particulate matter can be deposited 10 km of the point of emission. For the gaseous phase, deposition can take place 200–2,000 km from a source. The general short atmospheric residence times for the large fraction of airborne metals mean that changes in human activities contributing to emission to air are reflected fairly rapidly in local ambient concentrations (Brown et al., 1990). Hence a scaled analysis of heavy metals in soils impacted by urbanization is of importance.

10.4.3   Local-Scale Heavy Metal Soil Contamination

Purves (1972) and Purves and Mackenzie (1969) were among the first workers to recognize that soils in urban and industrial areas contain elevated levels of potentially toxic trace elements such as Cu, Pb, Zn, and B. Working in Edinburgh area, they found that the average urban soil contained more than 2 times as much available B, 5 times as much Cu, 17 times as much Pb, and 18 times as much Zn as those collected from adjacent rural areas (Purves, 1972; Gilbert, 1989). Since then several studies have been undertaken to assess anthropogenic sources of heavy metals.

Within the developed world, there are many examples of high levels of contamination within cities. For example, during the 1980s, in the UK, a national survey of house dusts in 53 representative villages, towns, and city boroughs revealed that in 10% of the 5,228 homes tested, the Pb concentration was in excess of 2,000 μg g^{-1} (Bullock & Gregory, 1991).

The concentrations of heavy metals (Cd, Cr, Cu, Hg, Ni, Pb, and Zn) and arsenic (As) were surveyed and the metal pools estimated in soils throughout Stockholm Municipality by Linde et al. (2001). Soils were sampled to a maximum depth of 25–60 cm. The results demonstrate a wide range in heavy metal concentrations, as well as in other soil properties. In the city center soils, concentration levels were homogeneous, whereas outside this area no geographical zones could be distinguished. The city center and wasteland soils generally had enhanced heavy metal concentrations to at least 30 cm depth compared to park soils outside the city center and rural (arable) soils in the region, which were used to estimate background levels. For example, the mean Hg concentration was 0.9 (max 3.3) mg kg^{-1} soil at 0–5 cm and 1.0 (max 2.9) at 30 cm depth in the city center soils, while the background level was 0.04 mg kg^{-1}. Corresponding values for Pb were 104 (max 444) and 135 (max 339) mg kg^{-1}, at 0–5 and 30 cm, respectively, while the background level was 17 mg kg^{-1}.

The average soil pools at 0–30 cm depth of Cu, Pb, and Zn were estimated as 21, 38, and 58 g m^{-2} respectively, which for Pb was 3–4 times higher and for Cu and Zn 1.5–2 times higher than the background level. The total amount of accumulated metals (down to 30 cm) in the city-center soils (4.5×10^6 m^2 public gardens and green areas) was estimated at 80, 1, 1, 120, and 40 t for Cu, Hg, Pb, and Zn, respectively.

Three conclusions were drawn from the study: (1) from a metal contamination point of view, more homogeneous soil groups were obtained based on present land use than on geographic distance to the city center; (2) it is important to establish a background level in order to quantify the degree of contamination; and (3) soil samples must be taken below the surface layer (and deeper than 30 cm) in order to quantify the accumulated metal pools in urban soils (Linde et al., 2001).

Manta et al.(2002) applied principal components and cluster analyses to investigate the sources of 11 heavy metals (Pb, Hg, Cu, Cr, Zn, Sb, Mn, V, Co, Ni, and Cd) in urban soils in Palermo (Sicily), Italy. Seventy topsoil (0–10 cm) samples were collected from green areas and public parks within the city. To account for the influence of lithology on heavy metal concentration, calcarenites, and cemented detritus samples were included for geochemical analysis. The results show that Hg, Pb, Zn, Cu, and Sb concentrations are higher for topsoils compared to their levels in unpolluted soils (at the world scale) and in the natural soils of Sicily. Cobalt, Cr, Ni, and Mn concentrations exhibit generally low levels, close to those reported for unpolluted soils. These metals also display quite homogeneous distributions across the city and therefore lower standard deviations, suggesting a major natural (i.e., indigenous lithologic) source. Cd and Mn values are comparable to values reported for natural soils worldwide and agree with those of natural soils in Sicily and those of rock samples collected in Favorita Park. Compared to average concentrations in urban soils, the median values of Pb in the analyzed soils are much lower than those reported for samples from some large and/or industrialized cities (i.e., Boston, central Madrid, central London), but they are similar to those measured in smaller cities (i.e., Hamburg, Glasgow) and residential areas of London (London Borough). Copper, Cr, Co, Zn, V, and Ni concentrations are generally similar to those reported for other cities, while Mn contents are generally higher (Manta et al., 2002).

The Manta et al. (2002) study suggests that heavy metal concentrations of urban soils can be used as geochemical tracers for monitoring the impact of human activity, provided that background levels have been correctly interpreted and established. Topsoil samples from green areas in the city of Palermo show Pb, Zn, Cu, Sb, and Hg concentrations higher than those of natural soils in Sicily and comparable to those recorded in other important European cities. Based on the whole dataset, these metals are inferred to derive from anthropogenic sources, whereas Co, Ni, V, Cr, and Mn distributions are mainly controlled by lithogenic inputs. Except for a few anomalously high Cd values that are an expression of the influence of pollutant sources, relatively high levels of Cd in the investigated soils were interpreted to reflect a natural enrichment by weathering and pedogenesis. Furthermore statistical analyses suggested that the most important source of the pollutant metals was vehicular traffic.

Soil samples collected at 10, 20, and 30 cm depths from 36 sites in Naples manifested a decrease in total content of C, Cr, Pb, and Zn with depth (Imperato et al., 2003). Moreover, high levels of all metals are found in sites of the eastern part of the city, corresponding with areas of heavy industry and where various oil refineries operate. Lead concentrations fluctuate throughout the city. The most contaminated soils are in the proximity of the motorway and streets with high traffic flows. Only soils on the northwest part of the city, which are characterized by greater elevations, contain low Pb levels. Despite the sharp increase of unleaded fuel utilization, followed by a rapid decline of Pb levels in the atmosphere, the content of Pb in urban soil still remains high with a consequent associated risk for children via the soil–hand–mouth pathway. About 14% of the overall analyzed soils showed levels of Cu, Pb, and Zn above the regulatory limits, forcing researchers to conclude that that the surface soils of Naples urban area appeared to be polluted in the order of Pb > Zn > Cu > Cr. Comparison between the 1999 data and those of 1974, revealed significant increases in Cu, Pb, and Zn contents of the surface layer of all the Naples urban soils (Imperato et al., 2003).

The influence of land use on metal contents of soils of Seville was studied by Ruiz-Cortes et al. (2005). Fifty-two samples corresponding to five categories of land uses: agricultural, parks, ornamental gardens, riverbanks, and roadsides were analyzed. Lower organic C, total N, and available P and K contents were found in riverbank samples, probably due to the lack of manuring of those sites, left in a natural status. Concentrations of Cu, Pb, and Zn were clearly higher in soils from ornamental gardens, whereas the concentrations in the riverbank samples were slightly lower than the other categories. In contrast, other metals (Cd, Cr, Fe, Mn, and Ni) were uniformly distributed throughout all land uses. A strong statistical association was observed among the concentrations of Cu, Pb, Zn, and organic C, suggesting that the larger contents of these metals in ornamental gardens are partly due to organic amendments added to those sites more frequently than to other kinds of sites.

Soils in developing world cities also demonstrate high levels of heavy metal contamination. For example, trace metal distribution of soils in the Danang-Hoian area (Vietnam) was studied by Thuy et al. (2000), who compared industrial, urban, rural, and cropland. Cu, Ni, Zn, and Zr show significant effects in industrial areas. Extremely high levels of Pb (up to $742 \mu g\ g^{-1}$) are observed in the industrial soil category, which shows an enrichment factor of 114 compared to rural soils. Cd shows only a relative local enrichment with a maximum level of $4.6 \mu g\ g^{-1}$ in urban soils (Thuy et al., 2000). Industrial soils how the highest values of Pb, Zn, Ni, and Cu. Pb concentrations vary between 130 and $740 \mu g\ g^{-1}$ with an average concentration of $327 \mu g\ g^{-1}$. Pb contents of industrial soils were 46, 91, and 20 times higher than in rural, urban, and crop soils, respectively. Other metals such as Zn, Zr, Cu, and Ni have similar distribution patterns. The highest concentrations of Zn (717 $\mu g\ g^{-1}$) and Ni ($240 \mu g\ g^{-1}$) were found in industrial soils. The elevation of Pb as well as other metal contents may be caused by the deposition of dust emitted and waste from various industries (Thuy et al., 2000).

Intensive urbanization of the Croatian capital Zagreb has led to the entrapment of good agricultural soils, developed mostly on Pleistocene aeolian sediments and alluvial and proluvian Holocene sediments within urban and suburban areas. Romic and Romic (2003) studied the influence of urban and industrialized environments on the accumulation of metals in these soils. A total of 331 samples were taken according to a regular 1×1 km square mesh over 860km^2. The following concentration ranges were observed: Cd 0.25–3.85 mg kg^{-1} (average 0.66 mg kg^{-1}); Cu 4.3–183 mg kg^{-1} (average 20.8 mg kg^{-1}); Fe 5.8–51.8 g kg^{-1} (average 27 mg kg^{-1}); Mn 79.2–1,282 mg kg^{-1} (average 613 mg kg^{-1}); Ni 0.70–282 mg kg^{-1} (average 49.5 mg kg^{-1}); Pb 1.50–139 mg kg^{-1} (average 25.9 mg kg^{-1}); and Zn 15.2–277 mg kg^{-1} (average 77.9 mg kg^{-1}). Factor analysis grouped Cd, Pb, Cu, Zn, and partially Ni into Factor 1 characterized with strongly scattered anthropogenic influence. The elements in Factor 2, Fe, Mn, and partially Ni are mainly of pedogenic origin—composition of different regolithic substrates of the fluvial origin in recent pedogenesis. High concentrations of Ni are also related to morphogenetic characteristics of the wider region, primarily basic and ultrabasic magmatic rocks of the surrounding mountain range. It was also suggested that the anomalous Ni concentrations in the vicinity of the highway and the airport were caused by fuel combustion. Copper is characterized by strongly scattered anthropogenic influence, which is related particularly to uncontrolled solid-waste disposals or discharges of liquid waste from households or agricultural enterprises. With Zn, Pb, and Cd, there are two possible ways to diffuse pollution. The Sava river, which drains the area and feeds the abundant Quaternary aquifer spreading below the major part of the investigated agricultural areas, have been exposed to intensive pollution by mining, industry, and cities in recent history. The part of the area with the highest determined concentrations of Zn, Pb, and Cd was repeatedly flooded as recently as the previous decade; therefore, the recent sedimentation of the river deposits exposed to pollution is a very probable cause of the accumulation of metals in this until recently inundation area. The other cause is atmospheric deposition of particles from urban sources (industrial emission, traffic, waste disposals, and heating plants). In addition to agricultural enterprises, several economically important, but ecologically risky, facilities are situated in the vicinity of the water-protection area. The area is intersected by a very busy ring road, while a marshalling yard, the city dump, pharmaceutical and chemical industry, the district-heating plant, and the airport are all located in close proximity.

Wang et al. (2005) also applied to principal components analysis to discriminate the sources of 30 heavy metals within the top 10 cm of Xuzhou soils in China. Three principal components accounted for over 70% variability in the data. The first principal component accounting for about 28% of the data showed a high loading with Al, Ti, Ga, Li, V, Co, Pt, Mn, and Be. Furthermore, the metals showed a high correlation ($P < 0.05$) with clay content ranging from 0.45 for Ga to 0.76 for Mn. This suggests that the variability of the metals is mainly controlled by soil parent material. Principal component II with high loadings on Ag, Se, Sc, Pb, Cu, Zn,

Cd, Au, Ni, S, and Mo and accounting for 24% variability was interpreted as anthropogenic factor connected with vehicular traffic. In particular, Zn compounds are employed as antioxidants in lubricating oils. The third principal component accounting for about 18% variability showed a high correlation with Bi, Cr, As, Hg, and Sb. It is related to the location of anthropogenic activities such as coal burning and similar industries in Xuzhou. Wang et al. (2005) further observed that Fe, Ba, Sn, Pd, and Br exhibited high loadings on both principal components I and II, suggesting that the metals are affected by both natural and anthropogenic sources.

A recent study in the Bangkok metropolitan region sampled 30 soils for heavy metal concentrations at 0–5 cm depth along a north–south bound "main axis" with "suburb", "central" and "industrial" branches defined at a right angle. All soils were Eutric or Dystric Gleysols derived from 2 to more than 30-year-old deposits consisting of clayey loam, quartz sand, and often high waste contents. In bulk soil and along the main axis also in aggregate core and surface fractions, Al, Cd, Cr, Cu, Fe, Mn, Ni, Pb, and Zn were sequentially extracted in seven fractions. At some sites the southern part of the study region Cd (up to 2.5 mg kg^{-1}), Cu (283), Pb (269), and Zn (813) concentration were high. Based on a principal component analysis the metal groups (i) Al, Cr, Fe, Mn, and Ni dominated by the concentration of the parent materials and (ii) Cd, Cu, Pb, and Zn dominated by anthropogenic input may be distinguished (Wilcke et al., 1998). This study also demonstrated that maximum total heavy metal concentrations in Bangkok soils were lower than in Manila, London, or Hamburg and except for Cd comparable to those of Hong Kong topsoil. There are, however, clearly contaminated sites which are concentrated on the southern part of the studied Bangkok metropolitan region (Wilcke et al., 1998).

In Hong Kong, urban parks are built close to major roads or industrial areas, where they are subject to many potential pollution sources, including vehicular exhausts and industrial emissions. Soil samples and associated street dusts were collected from more than 60 parks and public amenity areas in old urban districts, industrial areas and new towns of the territory. Soils were also sampled in the remote country parks to establish the baseline conditions. The total concentrations of heavy metals and major elements in the samples were determined. The results indicate that urban soils in Hong Kong have elevated concentrations of Cd, Cu, Pb, and Zn. The parks with high metal concentrations are located in old urban commercial districts and industrial areas, indicating that the major contamination sources in these soils are traffic emissions and industrial activities. In addition, the application of Cd-containing phosphate fertilizers may be an important source of Cd in urban park soils. The street dusts have highly elevated Zn concentration, particularly along the main trunk roads. The high Zn content in the street may come from traffic sources, especially vehicle tires.

Another study in Hong Kong by Chen et al. (1997) demonstrates that increases in trace metal concentrations in soils were generally extensive and obvious in urban and orchard soils, less in vegetable soils, whilst rural and forest soils were subjected to the least impact of anthropogenic sources of trace metals. However, some of the forest soils also contained elevated levels of As, Cu, and Pb. Urban soils in Hong Kong were heavily polluted by Pb from gasoline combustion. Agricultural soils, both orchard and vegetable soils, usually accumulated As, Cd, and Zn originating from applications of pesticides, animal manures, and fertilizers. In general, trace metal pollution in soils of the industrial areas and Pb pollution in the soils of the commercial and residential areas were obvious (Chen et al., 1997).

A recent study in India investigated heavy metal pollution of soils near Thane-Belapur industrial belt of Mumbai. Mumbai is a heavily populated industrial island city on the west coast of India, which is known as the commercial capital of the country. Fundamental to rapid urbanization are the numerous industries at Thane-Belapur industrial area. It houses 400 industries within 20 km area with all types of process industries, including chemical, pharmaceutical, textile, steel, paper, plastic, and fertilizers (Krishna & Govil, 2005). Soil samples collected from surrounding industrial areas were analyzed for toxic/heavy metals. The results indicate enrichment of the soils with Cu, Cr, Co, Ni, and Zn. The concentration ranges were: Cu 3.10–271.2 mg kg^{-1} (average 104.6 mg kg^{-1}); Cr 177.9–1,039 mg kg^{-1} (average 521.3 mg kg^{-1}); Co 44.8–101.6 mg kg^{-1} (average 68.7 mg kg^{-1}); Ni 64.4–537.8 mg kg^{-1} (average 183.6 mg kg^{-1}); and Zn 96.6–763.2 mg kg^{-1} (average 191.3 mg kg^{-1}) (Krishna & Govil, 2005).

Soils and dusts in two representative satellite cities (Uijeongby and Koyang) of Seoul, Korea demonstrated seasonal variations in metal concentrations through the rainy season. Concentrations of Cu, Pb, and Zn were higher than those of the world averages for soils, and their levels decreased after rain, particularly in highly contaminated samples. Relatively high pH values were found in roadside soils, but no seasonal variation was found after the rainy season. Dust samples had higher concentrations of Cu, Pb, and Zn than soil samples both before and after the rainy season. In terms of mobility and bioavailability of metals in soils and dusts, the order Zn > Cu > Pb is suggested. Geographical variations of total metals corresponded well with urbanized areas of cities, especially the industrial complex and major motorways.

Few studies are made on the potential soil Pb burden for small towns in rural environments. In one study, older homes were higher sources of higher levels of Pb, exceeding 1,000 μg g^{-1} than newer homes (Francek, 1992). The relationship, between home age and Pb level, was significantly positive ($r = 0.59$). Schools, which are mainly located away from heavily traveled roads and typically of brick construction have soil Pb concentrations at background levels. In general, the small city Pb burden is lower than in major urban areas. However, soils around older homes and in special locales, such as salvage yards, have Pb levels comparable to major urban areas.

While mean levels of Pb among different locations were weakly significant within small towns (Francek, 1992), median soil Pb levels were higher along the most heavily traveled roads (320 μg g^{-1}) than background concentrations (200 μg g^{-1}). This finding points to the importance of the intensity of activities within cities as a determinant of contamination. In another study, in Auckland, New Zealand, however, traffic flow as associated with larger differences of heavy metals. In areas were traffic was 50,000 vehicles or more per day, Pb levels were 2,200 μg g^{-1}, compared to a background level of 14 μg g^{-1}. The contamination levels have been related to components of gasoline, motor oil, car tires, and the roadside deposition or residues from these materials.

Another study suggests that distribution of heavy metal pools is associated with different types of land-use infrastructure (Sorme et al., 2001). Over a course of the century, within Stockholm, Sweden, pools of metals have accumulated differentially around infrastructure, buildings, homes, and enterprises. Data suggest that Cd, Cr, Cu, Hg, Ni, Pb, and Zn are associated differentially with various types of infrastructure.

Finally, another important factor in determining differences in soil heavy metal contamination include whether the soils were treated with sludge or manure. Those soils found to be pretreated with sludge or other amendments had higher levels of heavy metals than those that were not (Purves, 1972). At the same time, however, the nontreated urban soils still had higher levels of B, Cu, Pb, and Zn than rural soils.

In summary, soils in cities throughout the world are polluted with heavy metals. The distribution of these metals, however, varies by depth of soil and location around the city, largely dependent upon land use, intensity of human activity, infrastructure and pretreatment, and of course, weather and wind patterns. Sometimes, bedrock sources are important in determining high levels of metals within soils. Surprisingly, high levels of pollutants, even higher than those in the developed world, are found in developing world cities.

10.4.4   Metropolitan-Wide Distribution of Heavy Metals

Concentrations of heavy metals in urban topsoil vary considerably across different cities possibly due to differences in socioeconomic development and enforcement of environmental regulations. The question that this section addresses is whether the pattern of soil contamination differs with distance from a city center. Put another way, this section presents the regional patterns of soil contamination.

One of the early studies on soil contamination in England was carried out by Davies (1984). The study investigated heavy metal contamination with respect to distance from the city center of Birmingham. It was postulated that contamination processes known to affect city soils and vegetable would extent beyond the city limit to affect rural soils. Concentrations of Cd, Cu, Pb, and Zn in soils and plants along a 22 km transect extending form the city center, revealed a general order of contamination urban > suburban > rural. The distance decline relationships modeled using polynomial regression analysis and quadratic regressions yielded satisfactory fits. The initial hypothesis was confirmed arising from urban and industrial activities (Davies, 1984).

Judged by UK criteria (Pb = 65, Zn = 48, Cu = 16, and Cd = 1.6 mg kg^{-1} soil), Davies (1984) found that of the 31 soils analyzed throughout the urban region, 55% were contaminated by Pb, 68% by Zn, 71% by Cu, and 6% by Cd.

Lead concentrations were also measured in house dust, pavement dust, road dust, and garden soil in and around 97 inner-city houses in Birmingham, England (Davies, 1984;). The highest mean dust Pb concentration within the home, 615,

μg g^{-1}, is noted in samples under the doormat. Generally the house dust Pb levels were lower than the national mean (507 μg g^{-1}), although soil Pb concentrations were slightly higher. The age of the property was found to influence the Pb levels in both house dust and garden soil, with older houses (>35 years) having significantly higher concentrations than newer properties (<35 years). Houses being decorated at the time of sampling were found to have significantly higher Pb concentrations than those that were not. Elevated Pb levels were also noted in house dust and garden soil from houses located within a 500 m radius of commercial garages. Increased Pb concentrations were found in soil samples from gardens in close proximity to wastelands (demolition sites and tips), metal-using industries, and from those within 10 m of a road. Road dust samples from industrial areas had significantly higher Pb concentrations than those from residential areas.

In the USA, Carey et al. (1980) carried out studies of the heavy metals and pesticides found in lawns and wastes from five cities and compared these samples to those obtained in the respective suburban areas. Lawn sites are defined by the authors as mowed areas near houses and other structures, mowed parks, gardens, or cultivated areas, and obvious house yards. Waste sites are defined as abandoned lots, small wooded lots, power and gas lines, and bare soil exposed around construction sites, eroded areas, etc. The main source of the contamination appears to be from the deposition of metallic aerosols from industrial processes and/or the burning of fossil fuels. Lead has the highest concentrations followed by As in the urban areas. Results demonstrate that in almost every case there are higher levels of Pb, Hg, Cd, and Ar, in the inner city locations than in the suburban locations.

Burguera et al. (1988) investigated the Pb content of soil along the roadside in Merida city of Venezuela. Where motor vehicle traffic volume was <5,000 vehicles per day, almost no Pb accumulated was observed in the surface soil (0–2 cm), but where the motor vehicle traffic volume was >10,000 vehicles per day, levels of Pb increased by a factor of up about 18. The soil Pb content generally decreased with distance from the edge of the roadside and with depth of sampling.

Ordonez et al. (2003) characterized the elemental composition of street dust and soils in Aviles, northern Spain. Aviles is a medium-size city of approximately 80,000 inhabitants, where industrial activities and traffic strongly affect heavy metal distribution. Elevated mean concentrations of Zn (4, μg g^{-1}), Cd (22.3 μg g^{-1}), and Hg (2.56 μg g^{-1}) in street dust were found in industrial area samples. The researchers distinguished between two types of human influence. Metallurgical activity and transportation of raw materials for industries was the first and most important influence. The second includes exhaust emissions from traffic leading to elevated Pb concentration in areas with high vehicular density (mean 514 μg g^{-1}). The Zn content in the dust samples decreased with the distance from a Zn smelter located in the northern part of the city. The same trend was found for other elements in association with Zn in the raw materials used by the smelter, such as Cd and Hg. A simultaneous research campaign of urban soils, that involved the collection of 40 samples from a 10 km^2 area, revealed mean concentrations of 376 μg g^{-1} Zn, 2.16 μg g^{-1} Cd, 0.57 μg g^{-1} Hg, and 149 μg g^{-1} Pb. The distribution patterns were almost identical to those found for street dust (Ordonez et al., 2003).

In New York City, Pouyat (1991) initiated a study examining the nutrient cycling rates along an urban–rural gradient from the center of the city to the Connecticut countryside. Heavy metal concentrations were determined in the forest floor and the soil at nine sites along the gradient. The level of contamination in the urban stands is high, in contrast to the relative uniformity of Mn concentration among the sites. Manganese is important because it may reach toxic plant levels and interfere with uptake of other nutrients. The lack of clear trends for Co, Cr, and Cu indicates the importance in other factors in distribution patterns. While the accumulation of these pollutants in the forest floor should be expected, the high levels in the surface mineral soil are surprising. The author explains this translocation on earthworm activity incorporating the contaminant-containing organic material into the surface mineral soil (see also Pouyat et al., 1997).

In summary, these studies demonstrate that for some pollutants a rural–urban gradient is evident, while for others the trend is less clear. To a large extent differences are associated with the importance of air pollution, and specific industrial activities. Other factors, related to urban activities also play a role in the patterns described, but more work and different techniques are needed to uncover their relative influence.

10.4.5 Urbanization and Heavy Metal Contamination at the Regional and Global Scale

Urban areas have historically been the sources of regional and global pollutants. Cities are particularly influential in global emissions of fossil fuels (Organization for Economic Cooperation and Development, 1995). Studies have been carried out to assess the impact of fossil-fuel burning on agricultural lands further afield (Chameides et al., 1994). Chameides et al. (1994) reports that the most pronounced impact are in three regions of the northern midlatitudes: (i) North America, (ii) Europe, and (iii) eastern China and Japan. Within each of these regions, intense urban-industrial and agricultural activities tend to cluster together into a single large network, or plexus, of lands affected by human activity called CSMAPs. Certainly, "Giant Brown Clouds" or "Atmospheric Brown Clouds" are associated with urbanization.[13] Not only do these regions account for about 75% of the world's consumption of commercial energy and fertilizers, but they are also the major sources for atmospheric pollutant such as nitrogen oxides (NO_x), which along with ozone (O_3) and volatile organic compounds (VOC), are important precursors to photochemical smog refers.

High concentrations of O_3 in and around urban areas, combined with growing numbers of automobiles and expanding roadway networks and increasing reliance on nitrogenous fertilizers, has greatly increased the spatial scale of photochemical smog

---

[13] See for example http://www.nasa.gov/vision/earth/environment/brown_cloud.html.

in the CSMAPs. Regional O_3 pollution, often associated with summertime, can extend over thousands of kilometers and encompass agricultural as well as urban areas. The repetition of episodes of O_3 pollution over a growing season produced a pattern of chronic exposure that ultimately reduces crop yields (Chameides et al., 1994).

Simulations suggest that a sizable portion of the world's food crops are presently exposed to high NO_x emissions from CSMAPs. In fact, simulations for China suggest that as nonurban O_3 increases with industrialization, its effects on crops could hinder efforts to meet increasing food demands in the coming decades (Chameides et al., 1999).

10.5   Conclusion

This assessment of the impact of urbanization on soils is preliminary. The review only provided a cursory view of current trends and conditions globally. Notwithstanding the limitations, however, the study does demonstrate the dramatic impact that urbanization has on soils. Two particularly important areas demand further attention. First, cities and urban processes have had significant but varying impacts on soils structure, pH, and pollutant loads. One important conclusion of this review is that outside of very general conclusions, it is difficult to predict the impacts of urbanization on soils within any one location. Much depends upon both the unique conditions of the site and the direct and indirect factors creating change. This suggests the need for more studies throughout the world. It also suggests we need to perform a global assessment in this area to further identify common patterns and processes as well as unique circumstances.

Second, the study demonstrates that increasingly cities are impacting soils at different scales. That is, the power of urban activities to influence the biophysical and chemical conditions of soils at a distance is increasing. Given the growing importance of urban populations, this finding alone prompts us to further research and demands policy measures.

In terms of the specific implications of the findings for management and policy, we conclude that there is enough evidence of soil pollutant contamination in developing world cities at high-enough levels that warrant immediate and urgent action. Work on soil protection and remediation in these locations should be highly encouraged. This could be in through soil vapor extraction, chemical treatment, soil washing, biosorption, or phytoremediation. Of course, not all developing world cities have levels of pollution of concern. Responses will depend upon the industrial and increasingly mobile source levels, as well as the most efficient and appropriate form of remediation available. Of primary concern are those areas in the developing world that are growing through rapid industrialization.

We also conclude that there is much more work needed in this area. For one, this specific review warrants a full global assessment of the impact of urbanization on all environmental media. Given the importance of soils and the growing impacts identified in the study, a multiscale global assessment, similar to the Millennium Ecosystem Assessment for soils is desperately needed. Specific work that links urban activities to soils at a distance is a research area that has yet to emerge. Furthermore, we need to use and further develop new methods with which to study the impact of urban activities on soils. Certainly, current methods, such as urban–rural gradient analysis, has been helpful, but newer techniques, using GIS and satellite imagery, also promise to further our understanding of urban growth and soil quality and would focus on larger scales. Advances in these areas should therefore be promoted. Finally, studies of the impact of soils by urban activities can provide an important area for furthering interdisciplinary research (IDR). IDR is believed to be of increasing importance in environmental studies. We argue that studies of soils should include researchers from the social as well as physical sciences, particularly when identifying and analyzing forces that contribute to changes in soil biogeochemistry.

References

Ackerman, J. (2006). City parks: Space for the soul. *National Geographic, 210*, 110–115.

Agardy, T., Alder J., et al. (2005a). Coastal systems. In *Ecosystems and human well-being, Vol. 1: Current state and trends* (ed. M. E. Assessment). Washington, DC: Island Press.

Agardy, T., Alder J., et al. (2005b). Coastal systems. In R. Hassan, R. Scholes & N. Ash (Eds.), *Ecosystems and human well being: Current state and trends, Vol. 1*. Washington, DC: Island Press.

Berry, B. J. L. (1990). Urbanization. In B. L. Turner II, W. C. Clark, R. W. Kates, J. F. Richards, J. T. Mathews & W. B. Meyer (Eds.), *The earth as transformed by human action: Global and regional changes in the biosphere over the past 300 Years* (pp. 103–119). Cambridge: Cambridge University Press.

Berry, D. (1978). Effects of urbanization on agricultural activities. *Growth and Change, 9*, 2–8.

Blitzer, S., Hardoy, J., & Satterthwaite, D. (1981). Shelter: People's needs and governments' response. *Ekistics, 48*, 4–13.

Blume, H.-P. (1989). Classification of soils in urban agglomerations. *Catena, 16*, 269–275.

Bogue, D. J. (1956). *Metropolitan growth and conversion of land to non-agricultural uses*. Oxford University Press published jointly by Scripps, Oxford, Ohio.

Bogue, D.J. (1956). *Metropolitan growth and the conversion of land to non-agricultural uses*. Studies in population distribution, no. 11, Scripps Foundation, Oxford, OH.

Brady, N. C., & Weil, R. R. (2002). *The nature and properties of soils*. Upper Saddle River, NJ: Prentice Hall.

Braimoh, A. K. (2004). *Modeling land-use change in the Volta Basin of Ghana. Ecology and Development Series No. 14*. Bonn, Germany: University of Bonn, Center for Development Research.

Braimoh, A. K., Stein A., & Vlek, P. L. G. (2005). Identification and mapping of associations among soil variables. *Soil Science, 170*, 137–148.

Braimoh, A. K., & Vlek, P. L. G. (2006). Soil quality and other factors influencing maize yield in Northern Ghana. *Soil Use and Management, 22*, 165–171.

Brown, H. S., Kasperson, R. E., & Raymond, S. (1990). Trace pollutants. In I. B. L. Turner, W. C. Clark, R. W. Kates, J. F. Richards, J. T. Mathews & W. B. Meyer (Eds.), *The Earth as transformed by human action, global and regional changes in the biosphere over the past 300 Years* (pp. 437–454). New York: Cambridge University Press.

Bullock, P., & Gregory, P. J. (Eds.) (1991). *Soils in the urban environment*. Oxford: Blackwell.

Burguera, J. L., Burguera, M., & Rondon, C. (1988). Lead in roadside soils of Merida City Venezuela. *The Science of the Total Environment, 77*, 45–49.

Carey, A. E., Gowen, J. A., Forehand, T. J., Tai, H., & Wiersma, G. B. (1980). Heavy metal concentrations in soils of five United States cities, 1972 Urban Soils Monitoring Program. *Pesticides Monitoring Journal, 13*, 150–154.

Chameides, W. L., Kasibhatla, P. S., Yienger, J., & Levy, H. (1994). Growth of continental-scale metro-agro-plexes, regional ozone pollution, and world food production. *Science, 264*, 74–77.

Chameides, W. L., Li, X., Tang, X., Zhou, X., Luo, C., Kiang, C. S., St. John, J., Saylor, R. D., Liu, S. C., Lam, K. S., Wang, T., & Giorgi, F. (1999). Is ozone pollution affecting crop yields in China? *Geophysical Research Letters, 26*, 867–870.

Chen, T. B., Wong, J. W. C., Zhou, H. Y., & Wong, M. H. (1997). Assessment of trace metal distribution and contamination in surface soils of Hong Kong. *Environmental Pollution, 96*, 61–68.

Chhabra, R. (1985). *India: Environmental degradation, urban slums, political tension* (pp. 1–6). Draper Fund Report 14.

Craul P. J., & Klein C. J. (1980). Characterization of streetside soils of Syracuse, New York. *Metropolitan Tree Improvement Alliance (METRIA) Proceedings, 3*, 88–101.

Craul, P. J. (1985). A description of urban soils and their desired characteristics. *Journal of Arboriculture, 11*, 330–339.

Craul, P. J. (1992). *Urban soil in landscape design*. New York: Wiley.

Davies, B. E. (1984). Distance-decline patterns in heavy metal contamination of soils and plants in Birmingham, England. *Urban Ecology, 8*, 285–294.

Davis Kingsley (1965). "The urbanization of the human population" *Scientific American*.

De Kimpe, C., & Morel, J.-L. (2000). Urban soil management: A growing concern. *Soil Science, 165*, 31–40.

Decker, E. H., Elliott, S., Smith, F. A., Blake, D. R., & Rowland, F. S. (2000). Energy and material flow through the urban ecosystem. *Annual Review of Energy and Environment, 25*, 685–740.

Douglas, I. (1974). The impact of urbanization on river systems. *Proceedings of the International Geographical Union Regional Conference and Eight New Zealand Geography Conference*, 307–317.

Douglas, I. (1978). The impact of urbanization on fluvial geomorphology in the humid tropics. *Geo-Eco-Trop, 2*, 229–242.

Douglas, I. (1981). The city as an ecosystem. *Progress in Physical Geography, 5*, 315–367.

Douglas, I. (1983). *The urban environment*. London: Edward Arnold.

Douglas, I. (1994). Human settlements. In W. B. Meyer & B. L. Turner (Eds.), *Changes in land use and land cover: A global perspective* (pp. 149–169). Cambridge: Cambridge University Press.

Driessen, P., Deckers, J., & Spaargaren, O. (2001). Lecture notes on soils of the world soil resources reports, 94. Rome: FAO.

Ehrlich, A. (1991). Food and people. *Population and Environment: A Journal of Interdisciplinary Studies, 12*, 221–229.

European Environment Agency (2006) *Urban sprawl in Europe, The ignored challenge* (pp. 60). Copenhagen: EEA.

FAO. (1985). Urbanization: A growing challenge to agriculture and food systems in development countries. In FAO (Ed.), *State of food and agriculture 1984* (pp. 79–124). Rome: FAO.

FAO. (2006). *Global forest resources assessment 2005, Progress towards sustainable forest management*. Rome: FAO

Fischel, W. A. (1982). The urbanization of agricultural land: A review of the National Agricultural Lands study. *Land Economics, 58*, 236–258.

Folke, C., Jansson, A., Larsson, J., & Costanza, R. (1997). Ecosystem appropriation by cities. *Ambio, 27*, 167–172.

Francek, M. A. (1992). Soil lead levels in a small town environment: A case study from Mt. Pleasant, Michigan. *Environmental Pollution, 76*, 251–257.

Frey, H. T. (1984). Expansion of urban areas in the United States 1960–1980. Washington, DC: USDA Economic Research Service.

Garcia-Miragaya, J., Castro, S., & Paolini, J. (1981). Lead and zinc levels and chemical fractionation in roadside soils of Caracas, Venezuela. *Water, Air and Soil Pollution, 15*, 285–297.

Gardner, G. (1996). Asia is losing ground. *World Watch, 9*, 18–27.

Gbadegesin, A., & Olabode, M. A. (2000). Soil properties in the metropolitan region of Ibadan, Nigeria: Implications for the management of the urban environment of developing countries. *The Environmentalist, 20*, 205–214.

Gilbert, O. L. (1989). *The ecology of urban habitats*. London: Chapman & Hall.

Gill, D., & Bonnett, P. (1973). *Nature in the urban landscape: A study of city ecosystems*. Baltimore, MD: York Press.

Goffman, P. M., & Crawford, M. K. (2003). Denitrification potential in urban riparian zones. *Journal of Environmental Quality, 32*, 1144–1149.

Green, D. M., & Oleksyszyn, M. (2002). Enzyme activities and carbon dioxide flux in Sonoran desert urban ecosystem. *Soil Science Society of America Journal, 66*, 2002–2008.

Grubler, A. (1994). Technology. In W. B. Meyer & I. B. L. Turner (Eds.), *Changes in land use and land cover: A global perspective* (pp. 287–328). Cambridge: Cambridge University Press.

Hardoy, J. E., Mitlin, D., & Satterthwaite, D. (2001). *Environmental problems in an urbanization world*. London: Earthscan.

Hart, J. F. (1976). Urban encroachment on rural areas. *Geographical Review, 66*, 1–17.

Hassan, R., Scholes, R., & Ash, N (eds.), (2005) *Ecosystems and human well-being: Current state and trends, Vol. 1*. Washington, DC: Island Press.

Hillel, D. (1980). *Fundamentals of soil physics*. New York: Wiley.

Hiller, D. A. (2000). Properties of Urbic Anthrosols from an abandoned shunting yard in the Ruhr area, Germany. *Catena, 39*, 245–266.

Hooke, R. L. (2000). On the history of humans as geomorphic agents. *Geology, 28*, 843–846.

Hough, M. (2004). *Cities and natural process: A basis for sustainability*. (2nd ed.) London: Routledge, Taylor & Francis.

Huang, J., Lu, X. X., & Sellers, J. M. (2006). Urban form in the developed and developing worlds: An analysis using spatial metrics and remote sensing. *Environment and Urban Planning*.

Huang, J., Lu, X. X., & Sellers, J. M. (2007). A global comparative analysis of urban form: applying spatial metrics and remote sensing. *Landscape and Urban Planning, 82(4)*, 184–197.

Imperato, M., Adamo, P., Naimo, D., Arienzo, M., Stanzione, D., & Violante, P. (2003). Spatial distribution of heavy metals in urban soils of Naples city (Italy). *Environmental Pollution, 124*, 247–256.

Jacoby, H. J. (2000). Access to markets and the benefits of rural roads. *The Economic Journal, 110(465)*, 713–737.

Jim, C. Y. (1989). Tree-canopy characteristics and urban development in Hong Kong. *Geographical Review, 79*, 210–225.

Jim, C. Y. (1998a). Physical and chemical properties of a Hong Kong roadside soil in relation to urban tree growth. *Urban Ecosystems, 2*, 171–181.

Jim, C. Y. (1998b). Soil characteristics and management in an urban park in Hong Kong. *Environmental Management, 22*, 683–695.

Jim, C. Y. (2003). Soil recovery from human disturbance in tropical woodlands in Hong Kong. *Catena*, 85–103.

Kaya, S., & Curran, P. J. (2006). Monitoring urban growth on the European side of Istanbul metropolitan area: A case study. *International Journal of Applied Earth Observation and Geoinformation, 8*, 18–25.

Kelly, P. F. (1998). The politics of urban-rural relations: Land use conversion in the Philippines. *Environment and Urbanization, 10*, 35–54.

Kostel-Huges, F., Young, T. P., & Carreiro, M. M. (1998). Forest leaf litter quantity and seedling occurrence along an urban-rural gradient. *Urban Ecosystems, 2*, 263–278.

Krishna, A. K., & Govil, P. K. (2005). Heavy metal distribution and contamination in soils of Thane-Belapur industrial development area, Mumbai, Western India. *Environmental Geology, 47*, 1054–1061.

Landsberg, H. (1981). *The urban climate*. International Geophysics Series 28, New York.

Larson, W. E., Pierce, F. J., & Dowdy, R. H. (1983). The threat of soil erosion to long-term crop production. *Science, 219,* 458–465.

Lemay, M., & Mulamootil, G. (1984) A study of changing land uses in and around Toronto water-front marshes. *Urban Ecology, 8,* 313–328.

Li, X., Lee, S.-l., Wong, S.-c., Shi, W., & Thornton, I. (2004). The study of metal contamination in urban soils of Hong Kong using a GIS-based approach. *Environmental Pollution, 129,* 113–124.

Linde, M., Bengtsson, H., & Oborn, I. (2001). Concentrations and pools of heavy metals in urban soils in Stockholm, Sweden. *Water, Air and Soil Pollution: Focus, 1,* 83–101.

Lo, F.-c., & Marcotullio, P. J. (2000). Globalization and urban transformations in the Asia Pacific region: A review. *Urban Studies, 37,* 77–111.

Lorenz, K. & Kandeler, E. (2005). Biochemical characterization of urban soil profiles from Stuttgart, Germany. *Soil Biol. Biochem. 37,* 1373–1385.

Lubowski, R. N., Vesterby, M., Bucholtz, S., Baez, A., & Roberts, M. J. (2006). Major uses of land in the United States, 2002. Washington, DC: USDA, Economic Research Service.

Manta, D. S., Angelone, M., Bellanca, A., Neri, R., & Sprovieri, M. (2002). Heavy metals in urban soils: A case study from the city of Palermo (Sicily), Italy. *The Science of the Total Environment, 300,* 229–243.

McDonnell, M. J., Pickett, S. T. A., Groffman, P., Bohlen, P., Pouyat, R. V., Zipperer, W. C., Parmelee, R. W., Carreiro, M. M., & Medley, K. (1997). Ecosystem processes along an urban-to-rural gradient. *Urban Ecosystems, 1,* 21–36.

McGranahan, G., Jacobi, P., Songsore, J., Surjadi, C., & Kjellen, M. (2001). *The citizens at risk: From urban sanitation to sustainable cities.* London: Earthscan.

McGranahan, G., Marcotullio, P. J., et al. (2005) Urban systems. In *Current state and trends: Findings of the condition and trends working group. Ecosystems and human well-being, Vol. 1* (ed. Millennium Ecosystem Assessment, pp. 795–825). Washington, DC: Island Press.

McKinney, M. L. (2002). Urbanization, biodiversity and conservation. *BioScience, 52*(10), 883–890.

Newcombe, K., Kalma, J. D., & Aston, A. R. (1978). The metabolism of a city: The case of Hong Kong. *Ambio, 7,* 3–15.

Nizeyimana, E. L., Petersen, G. W., Imhoff, M. L., Sinclair, H. R., Waltman, S. W., Reed-Margetan, D. S., Levine, E. R., & Russo, J. M. (2001). Assessing the impact of land conversion to urban use on soils with different productivity levels in the USA. *Soil Science Society of America Journal, 65,* 391–402.

Nowak, D. J., Rowantree, R. A., McPherson, E. G., Sisinni, S. M., Kerkmann, E. R., & Stevens J. C. (1996). Measuring and analyzing urban tree cover. *Landscape and Urban Planning, 36,* 49–57.

Ordonez, A., Loredo, J., Miguel, E. D., & Charlesworth, S. (2003). Distribution of heavy metals in the street dusts and soils of an industrial city in northern Spain. *Archives of Environmental Contamination and Toxicology, 44,* 160–170.

Organization for Economic Cooperation and Development. (1995) *Urban energy handbook: Good local practice.* Paris: OECD.

Pavao-Zuckerman, M. A., & Coleman, D. C. (2007). Urbanization alters the functional composition, but not taxonomic diversity of the soil nematode community. *Applied Soil Ecology, 35,* 329–339.

Patterson, J. C. (1976). Soil compaction and its effects upon urban vegetation. In: *Better Trees for Metropolitan Landscapes Symposium Proceedings,* USDA Forest Service General Technical Report NE-22.

Pimentel, D., Harvey, C., Resosudarmo, P., Sinclair, K., Kurz, D., McNair, M., Crist, S., Shpritz, L., Fitton, L., Saffouri, R., & Blair, R. (1995). Environmental and economic costs of soil erosion and conservation benefits. *Science, 267,* 1117–1123.

Plaut, T. R. (1980). Urban expansion and the loss of farmland in the United States: Implications for the future. *American Agricultural Economics Association, 62,* 537–542.

Pouyat, R. V. (1991). The urban-rural gradient: an opportunity to better understand human impacts on forest soils. *Proceedings of the Society of American Foresters.* 1990 Annual Convention,

July 27–August 1, 1990, Washington, D.C., pp. 212–218. Society of American Foresters, Bethesda, Maryland.

Pouyat, R. V., Parmelee, R. W., & Carreiro, M. M. (1994). Environmental effects of forest soil-invertebrate and fungal densities in oak stand along an urban-rural land use gradient. *Pedobiologia 38*, 385–399.

Pouyat, R. V., McDonnell, M. J., & Pickett, S. T. A. (1997). Litter decomposition and nitrogen mineralization in oak stands along an urban-rural land use gradient. *Urban Ecosystems, 1*, 117–131.

Purves, D. (1972). Consequences of trace-element contamination of soils. *Environmental Pollution, 3*, 17–24.

Purves, D., & Mackenzie, E. J. (1969). Trace-element contamination of parklands in urban areas. *Journal of Soil Science, 20*, 288–290.

Randolph, J. (2004). *Environmental land use planning and management*. Washington, DC: Island Press.

Rees, W. E. (1992). Ecological footprints and appropriated carrying capacity: What urban economies leave out. *Environment and Urbanization, 4*, 121–130.

Rees, W. E. (2002). Globalization and sustainability: Conflict or convergence? *Bulletin of Science, Technology and Society, 22*, 249–268.

Rerat, A., & Kaushik, S. J. (1995). Nutrition, animal production and the environment. *Water Science & Technology, 31*, 1–19.

Rickson, R. J. (2003). Erosion risk assessment on disturbed and reclaimed land. In H. M. Moore, H. R. Fox & S. Elliott (Eds.), *Land reclamation, extending the boundaries* (pp. 185–192). Lisse, The Netherlands: A. A. Balkema.

Romic, M., & Romic, D. 2003. Heavy metals distribution in agricultural topsoils in urban area. *Environmental Geology, 43*, 795–805.

Rowntree, R. A. (1984). Forest canopy cover and land use in four eastern United States cities. *Urban Ecology, 8*, 55–67.

Rozanov, B. G., Targulian, V., & Orlov, D. S. (1990). Soils. In I. B. L. Turner, II, W. C. Clark, R. W. Kates, J. F. Richards, J. T. Mathews & W. B. Meyer (Eds.), *The Earth as transformed by human action, global and regional changes in the biosphere over the past 300 Years* (pp. 203–214). Cambridge: Cambridge University Press.

Ruiz-Cortes, E., Reinoso, R., Dias-Barrientos, E., & Madrid, L. (2005). Concentrations of potentially toxic metals in urban soils of Seville: Relationship with different land uses. *environmental Geochemistry and Health, 27*, 465–474.

Sanchez-Camazano, M., Sanchez-Martin, M. J., & Lorenzo, L. F. (1994). Lead and cadmium in soils and vegetables from urban gardens of Salamanca (Spain). The Science of the Total Environment, *146/147*, 163–168.

Sanders, R. A., & Stevens, J. C. (1984). Urban forest of Dayton, Ohio: A preliminary assessment. *Urban Ecology, 8*, 91–98.

Savage, V. R. (2006). Ecology matters: Sustainable development in Southeast Asia. *Sustainability Science, 1*.

Schleuss, U., Wu Q., & Blume H.-P. (1998). Variability of soils in urban and periurban areas in Northern Germany. *Catena, 33*, 255–270.

Schneider, A., Seto, K. C., Webster, D. R., Cai, J., & Luo, B. (2003). *Spatial and temporal patterns of urban dynamics in Chengdu, 1975–2002*. APARC: Stanford University.

Schueler, T. R. (1994). The importance of imperviousness. *Watershed Protection Techniques, 1*, 100–111.

Schulz, N. (2005). Contributions of material and energy flow accounting to urban ecosystems analysis: Case study Singapore. UNU-IAS Working Paper #136, Yokohama: UNU-IAS.

Short, J. R., Fanning, D. S., Foss, J. E., & Patterson, J. C. (1986a). Soils in the Mall in Washington, DC: II. Genesis, classification and mapping. *Soil Science Society of America Journal, 50*, 705–710.

Short, J. R., Fanning, D. S., McIntosh, M. S., Foss, J. E., & Patterson, J. C. (1986b). Soils in the Mall in Washington, DC: I. Statistical summary of physical properties. *Soil Science Society of America Journal, 50*, 699–705.

Spirn, A. W. (1984). *The granite garden*. New York: Basic Books.

Sorme, L., Bergback, B., & Lohm, U. (2001). Century perspective of heavy metal use in urban areas. *Water, Air and soil Pollution: Focus, 1*, 197–211.

Sukopp, H., Blume, H.-P., & Kunick, W. (1979). The soil, flora, and vegetation of Berlin's waste lands. In: *Nature in Cities, The Natural Environment in the Design and Development of Urban Green Space* (ed. I. C. Laurie) pp. 115–132. John Wiley & Sons, Chichester, UK.

Thuy, H. T. T., Tobschall, H. J., & An, P. V. (2000). Distribution of heavy metals in urban soils: A case study of Danang-Hoian Area (Vietnam). *Environmental Geology, 39*, 603–610.

Torrey, B. B. (2004). Urbanization: An environmental force to be reckoned with (pp. 6). New York: Population Reference Bureau.

UNESCAP. (2000). *State of the environment in the Asia Pacific*. Bangkok: UNESCAP.

United Nations. (1999). *World Urbanization Prospects: 1998 Revisions*. New York: DESA, UN.

United Nations. (2006). *World urbanization prospects: 2006 revisions*. New York: DESA, UN.

US Natural Resources Conservation Service. (1986). *Urban hydrology for small watersheds* (pp. 164). Washington, DC: United States Department of Agriculture, Conservation Engineering Division.

USDA. (1997). *Agricultural resources and environmental indicators 1996–97* (pp. 350). Washington, DC: Economic Research Service, Natural Resources and Environment Division.

USDA. (2001). *National resources inventory*. Washington, DC: Government Printing Office.

USDI. (2001). *Status and trends of wetlands in the coterminous United States, 1986 to 1997*. Washington, DC: US Department of Interior.

Vesterby, M., & Heimlich, R. E. (1991). Land use and demographic change: Results from fast-growth counties. *Land Economics, 67*, 279–291.

Vesterby, M., & Krupa, K. S. (2001). *Major uses of land in the United States, 1997* (pp. 47). Washington, DC: Resource Economics Division, Economic Research Service, US Department of Agriculture.

Wackernagel, M., & Rees, W. (1996). *Our ecological footprint*. Gabriola Island, Canada: New Society Publishers.

Wackernagel, M., Schulz, N. B., Deumling, D., Linares, A. C., Jenkins, M., Kapos, V., Monfreda, C., Loh, J., Myers, N., Norgaard, R., & Randers, J. (2002). Tracking ecological overshoot of the human economy. *Proceedings of the National Academy of Sciences of the United States of America, 99*, 9266–9271.

Warner, J. W., Jr. & W. E. Hanna. 1982. Soil Survey of Central Park, New York. Unpublished report, *Soil Conserv. Serv.*, USDA.

Wang, X. S., Qin, & S. X., Sang, (2005). Accumulation and sources of heavy metals in urban topsoils: a case study from the city of Xuzhou. *China Environ Geo., 48*, 101–107.

Wernick, I. K., & Irwin, F. H. (2005). *Material flows accounts: A tool for making environmental policy*. Washington, DC: World Resources Institute.

Wilcke, W., Muller, S., Kanchanakool, N., & Zech, W. (1998). Urban soil contamination in Bangkok: Heavy metal and aluminum partitioning in topsoils. *Geoderma, 86*, 211–228.

Wilkinson, B. H. (2005). Humans as geologic agents: A deep-time perspective. *Geology, 33*.

Wolman, A. (1965). The metabolism of cities. In *Cities* (ed. S. American, pp. 156–174). New York: Alfred A. Knopf.

Wu, J., & Hobbs, R. (2002). Key issues and research priorities in landscape ecology: An idiosyncratic synthesis. *Landscape Ecology, 17*, 355–365.

Yokohari, M., Takeuchi, K., Watanabe, T., & Yokota, S. (2000). Beyond greenbelts and zoning: A new planning concept for the environment of Asian mega-cities. *Landscape and Urban Planning, 47*, 159–171.

Young, I. M., & Crawford, J. W. (2004). Interactions and self-organization in the soil-microbe complex. *Science, 304*(5677), 1634–1637.

Index

CPSIA information can be obtained
at www.ICGtesting.com
Printed in the USA
LVHW080053091219
639845LV00002B/57/P